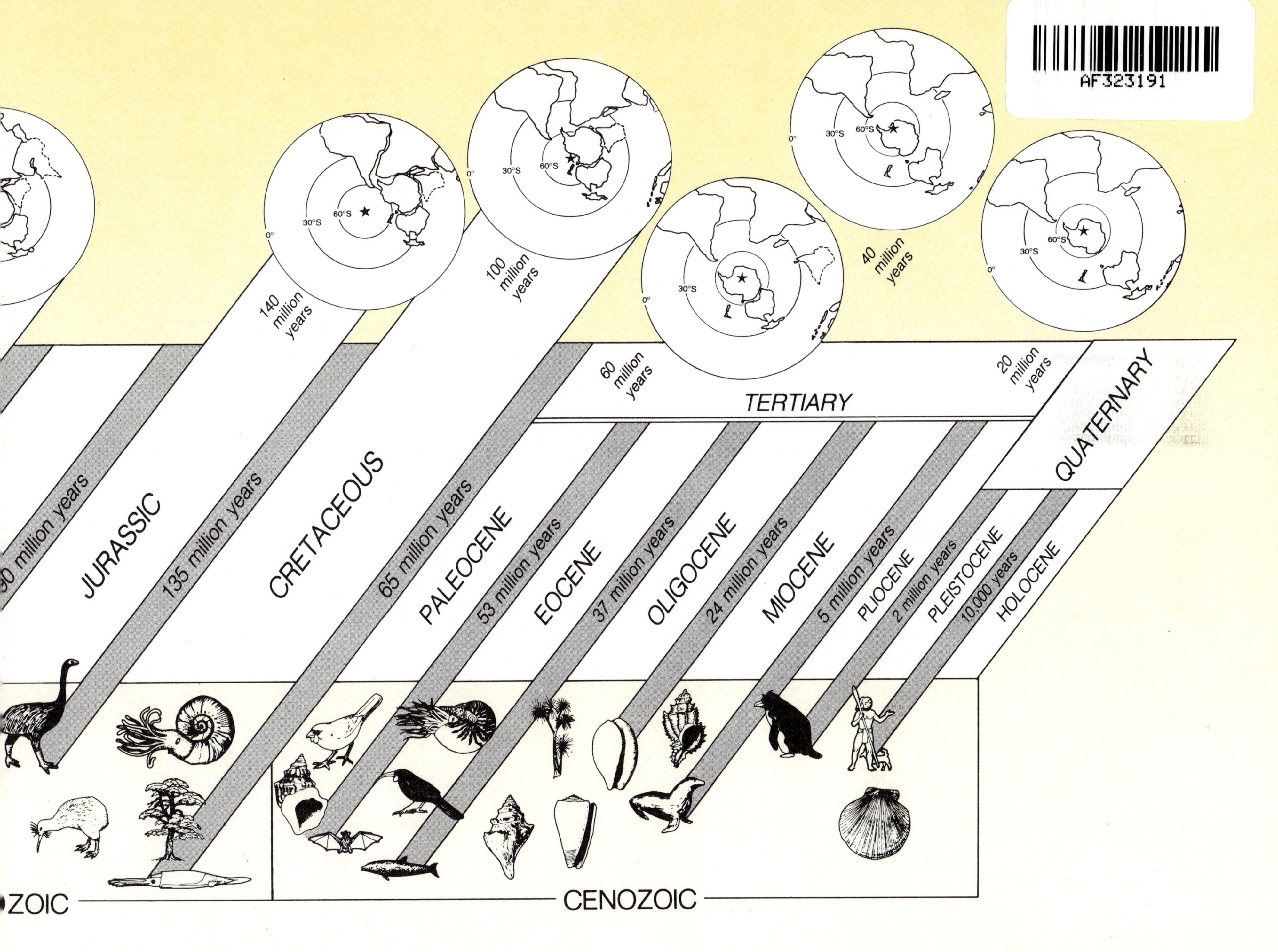
AF323191
0°
30°S
60°S
140 million years
100 million years
60 million years
40 million years
20 million years
0°
30°S
60°S
0°
30°S
60°S
0°
30°S
60°S
0°
30°S
60°S
0°
30°S
60°S
TERTIARY
QUATERNARY
90 million years
JURASSIC
135 million years
CRETACEOUS
65 million years
PALEOCENE
53 million years
EOCENE
37 million years
OLIGOCENE
24 million years
MIOCENE
5 million years
PLIOCENE
2 million years
PLEISTOCENE
10,000 years
HOLOCENE
ZOIC
CENOZOIC

LANDS IN COLLISION

DISCOVERING NEW ZEALAND'S PAST GEOGRAPHY

LANDS IN COLLISION

DISCOVERING NEW ZEALAND'S PAST GEOGRAPHY

by Graeme Stevens

New Zealand Geological Survey,   Department of Scientific and Industrial Research

SCIENCE INFORMATION PUBLISHING CENTRE   Wellington
1985

ISBN 0-477-06760-3

Edited for publication by Ian Mackenzie, SIPC, DSIR

DSIR Information Series No. 161

Published by Science Information Publishing Centre, DSIR,
P.O. Box 9741, Wellington, New Zealand

CATALOGUING IN PUBLICATION:

STEVENS, Graeme
    Lands in collision : discovering New Zealand's past
geography / by Graeme Stevens. – Wellington : SIPC,
1985.
    (DSIR information series, ISSN 0077-9636 ; no. 161)
        ISBN 0-477-06760-3
        UDC 551.248(931):551.7

COVER

Between about 140 and 110 million years ago earth movements associated with the opening of the Tasman Sea and Southern Ocean formed an ancestral New Zealand landmass. Erosion of the landmass began as soon as it was pushed up above sea level. The erosion proceeded at an inexorable pace, and by about 35 million years ago the ancestral landmass had been worn down and in most areas covered by the sea.

Renewed earth movements, beginning about 15–12 million years ago, disrupted the eroded remnants of the ancestral landmass and raised them to various heights above sea level. In many areas the layers of marine deposits laid down by the sea across ancestral New Zealand have been stripped off to reveal the underlying old worn-down land surface.

This aerial photograph looks northwards across the old land surface in the Poolburn–Rough Ridge area of Central Otago, east of Cromwell. Here the old surface has been eroded across hard schist rocks, residual knobs of which litter the terrain in the foreground.

The flatness of the old eroded land surface has been largely preserved in this part of Central Otago, but its original continuity has been partially interrupted by lines of fractures (fault lines) along which blocks of country have been tilted, raised, or lowered, forming longitudinal trough-like valleys and intervening block-like ranges (e.g., Knobby Range, Rock and Pillar Range, Raggedy Range, and Rough Ridge).

Although the old landsurface was formerly virtually continuous across New Zealand, large areas have been destroyed by erosion. The fragments present in Central Otago are perhaps the most continuous and the best preserved, but noteworthy fragments also occur in the Wellington region, and the Aorere valley (northwest Nelson).

NZGS photograph by Lloyd Homer

CONTENTS

REVOLUTION IN THE EARTH SCIENCES

The theory of continental drift and its modern version plate tectonics have given new insights into the processes and structure of our planet. The theory provides a unifying concept for the many facets of earth science.

These ideas have focussed attention on many features of the New Zealand region. The result has been a wealth of fresh information on how New Zealand evolved throughout geological time; how its distinctive geography was formed, and how it came to be the home of so many unusual native plants and animals.

In the late 1800s and early 1900s, as scientific exploration of the world progressed, it became more and more apparent that widely separated land masses had a number of things in common. They had the same underlying geological structure, and often shared the same rock layers and fossils. Modern plants and animals frequently showed close relationships. However, until the late 1950s and early 1960s it was fashionable to explain such geological, zoological, and botanical similarities between land masses by means of "sunken continents", or "land bridges", now submerged out of sight beneath the sea. The oceans and continents were regarded as essentially permanent and fixed in position. Following from this, the oceans were regarded as being very old, and were thought to date back to the formation of the earth. According to one theory, the Pacific Ocean was regarded as the scar left by the tearing away of a large part of the earth's surface to form the moon.

It was thought that eventually the land bridges sank beneath the sea. There grew up a tradition of sunken continents, assisted no doubt by tales of vanished lands handed down from Greek and Roman times (for example "Atlantis"). "Lemuria", for example, was the lost land that formerly linked Madagascar and India, allowing the monkey-like lemurs to reach both countries. Closer to home, "Tasmantis" was the vanished land on the site of the modern Tasman Sea. It was thought to have linked Australia and New Zealand, and allowed them to share animals and plants in past geological ages. There was an alternative to "land bridges", and that was to move the continents close together, so that animals and plants could move between them. Although such an alternative had been in some people's minds for some considerable time, it had been roundly condemned by many, and dismissed as geological heresy. It was regarded in the same way as Martians and flying saucers, and thought to be equally laughable.

THE BEGINNINGS

The general conformity in outline of South America and Africa had been speculated on by the philospher Francis Bacon (1620), the moralist Francois Placet (1666), and the naturalists Comte de Buffon (1778) and Alexander von Humboldt (1801). It was left, however, to the American/Italian Antonio Snider-Pellegrini to put forward the idea in 1858 that the coastlines of eastern South America and western Africa may fit together like a jig-saw puzzle. He formulated and published the first coherent ideas of what was later to become the theory of continental drift (Fig. 1).

These ideas were taken up and expanded by Eliseé Reclus in France (1872), F. B. Taylor in U.S.A. (1910), and more particularly by Alfred Wegener in Germany who published a series of books and papers between 1912 and 1924.

Wegener, a physicist and meteorologist by training, was particularly impressed by the jig-saw puzzle fit along both coasts of the Atlantic. He saw how the geology of countries like Norway, Spitzbergen, and Greenland matched, although now separated by vast expanses of the North Atlantic. By reading geological accounts published by other workers he was also able to appreciate that similar matches could be made elsewhere in the world. He realised that lands had a "geological grain", rather like lines of print in a newspaper and, in the same way as a torn newspaper may be put together using the lines of print as a guide, so geological structure can be used to match up individual lands that had formerly been together, but are now separated by expanses of ocean.

Wegener proposed that all the landmasses of the earth were once parts of a single continent called Pangaea, coined from the Greek and meaning "all the earth". Pangaea occupied half the earth's surface, the other half being covered by the primaeval Pacific Ocean. He envisaged that the super continent had gradually split up, opening gaps that became the modern oceans— a process of "continental drift" (Fig. 2).

Later, in the 1920s and 1930s, Wegener's ideas were elaborated upon by other workers, notably Alexander du Toit (South Africa) and Arthur Holmes (England). Both workers provided detailed geological and geophysical support for continental drift. However, du Toit differed from Wegener in that he postulated the existence of two super-continents, not one (Pangaea) as envisaged by Wegener. According to du Toit the Southern Hemisphere continents, with the addition of India, formed a southern landmass called Gondwanaland and the remaining continents a northern Laurasia (Fig. 3). The name Gondwanaland was derived from that of the Gonds, an ancient tribe in North India. The name Laurasia is a

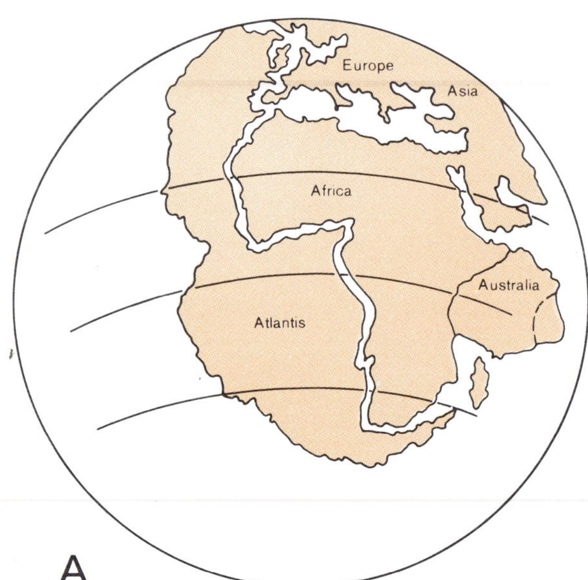

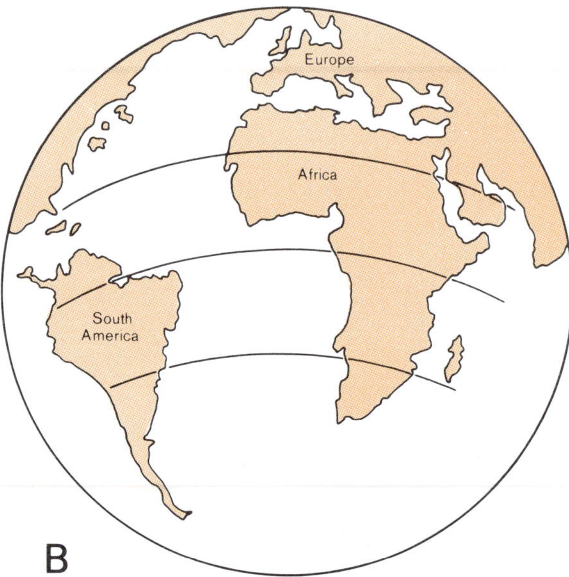

Fig. 1 Diagrams published by Antonio Snider in 1858 to illustrate his concept of the evolution of the world. Diagram (A) is the world as it appeared "Before the Separation", in the interval from Adam to Noah. Diagram (B) is the world "After the Separation", which took place during the Deluge.

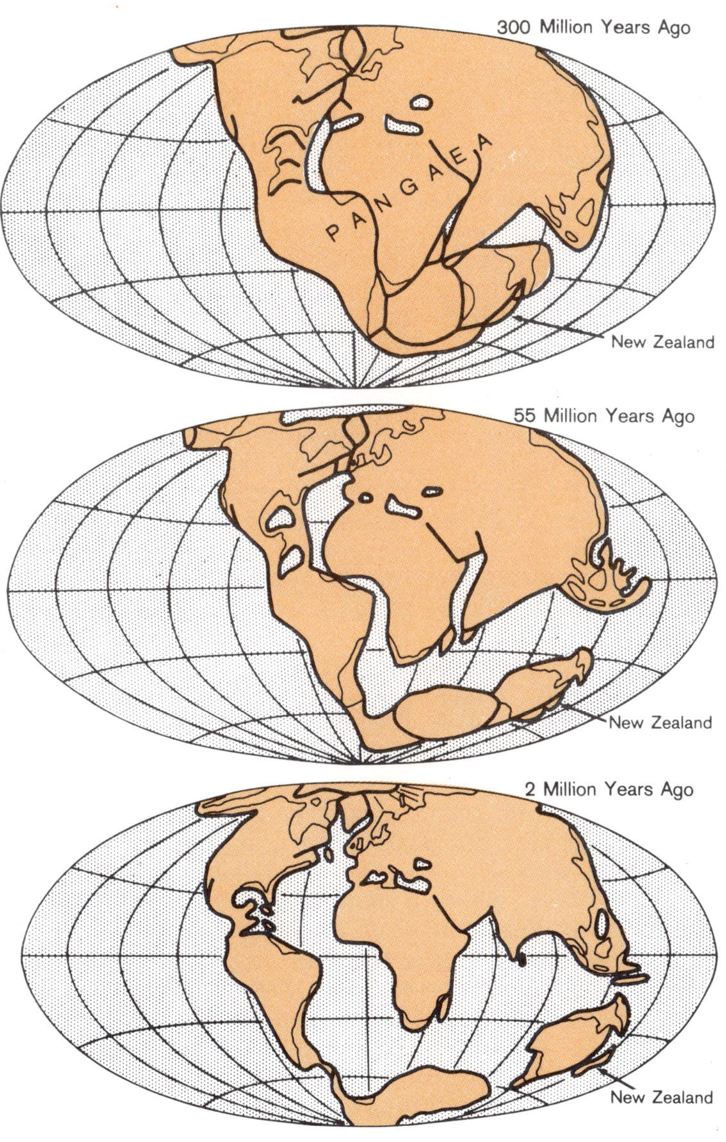

Fig. 2 Diagrams published by Alfred Wegener in 1929 to illustrate his concept of "Continental Drift", spanning the late Carboniferous (300 million years ago), Eocene (55 million years ago) and Quaternary (2 million years ago).

Fig. 3 Alexander du Toit introduced the concept of two super-continents or conglomerations of land: Laurasia and Gondwanaland, separated by a primaeval ocean called the Tethys. At various times the waters of the Tethys lapped across the edges of southern Europe and Asia, northern and eastern Africa, Arabia and India and branches of it sometimes extended to Australia, New Guinea, New Caledonia, New Zealand, and Antarctica, providing migration routes for shallow-water marine organisms.

As global movements brought Laurasia and Gondwanaland into collision, the old sea-floor sediments of the Tethys were buckled and rucked up to form the chains of Alpine mountains stretching from the Atlantic to Southeast Asia (Atlas, Rif, Betic, French Alps, Swiss Alps, Austrian Alps, Carpathians, Dinarides, Caucasus, Zagros, Himalaya, etc.). The modern Mediterranean Sea is a shrunken remnant of the once-extensive Tethys.

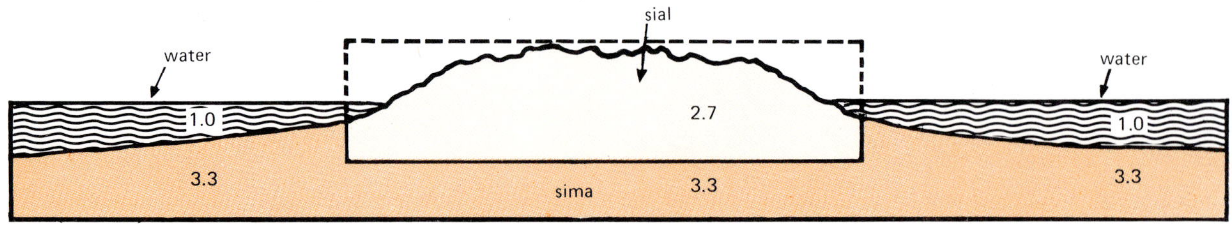

Fig. 4 Sketch to show differences in density between rocks of the continents (collectively known as "sial") and oceans (collectively "sima").

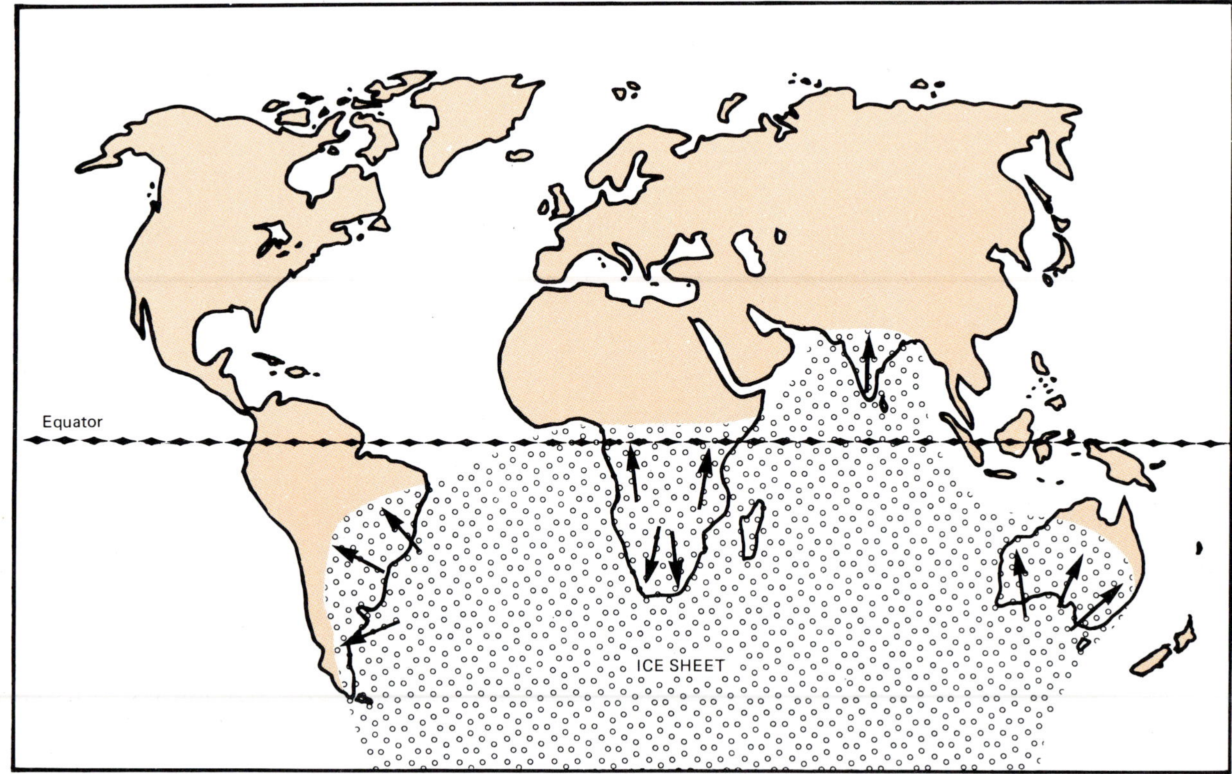

Fig. 5 The distribution of remnants of the Carboniferous–Permian glaciation (350–230 million years ago) is shown in relation to present-day geography. The directions of ice movement recorded on the various continents are shown by the arrows. The hypothetical ice sheet that would be required to glaciate the landmasses in the way indicated is depicted by the circle pattern. As can be seen, such an ice sheet, if it was indeed present, would have had to span most of the South Atlantic and Indian oceans (compare with Fig. 6).

combination of Laurentia (a geographic term for the region adjacent to the St Lawrence River), and Eurasia. du Toit maintained that the two super-continents had different and largely independent histories. From at least 250 million years ago the two super-continents were separated by an extensive sea, called the Tethys (named after the Greek goddess of the sea). This sea was not disrupted until 40 or 50 million years ago, as a result of Africa and India breaking away from Gondwanaland and drifting northwards towards Eurasia.

Wegener also highlighted a major flaw in the then current ideas involving land bridges and sunken continents. He pointed out that there is a fundamental difference in density between continental and oceanic rocks. If water is assumed to have a density of 1.0, then continental rocks have an average density of 2.7 whereas those of the ocean floors have an average density of 3.3 (Fig. 4). This situation is comparable to that of a cork floating in water. The cork will float, come what may, and can only be submerged by physically pushing it under with a finger or hand. However, once the finger or hand is removed, the cork pops up to the surface again. Wegener reasoned that it was not possible to simply submerge land bridges or lost continents such as Atlantis, Lemuria, or Tasmantis, without some type of heavy loading, analogous to the finger or hand keeping the cork submerged. Thereupon the advocates of a fixed earth responded by paring down the lost continents and land bridges to thin "isthmian links", rather like the modern Panama isthmus connecting North and South America. However, such a subterfuge merely lessened, but did not eliminate, the density problem.

Another way around the problem was the idea that land bridges could be loaded with quantities of dense volcanic and igneous rock— the former poured out from volcanoes, and the latter injected into the rocks from below. However, this idea did not find general acceptance.

In another masterly move, Wegener drew attention to the remnants of ice ages that occurred in the Carboniferous and Permian periods 350–230 million years ago. Such remnants are found today in Australia, Antarctica, Brazil and Argentina, southern Africa, and India (Fig. 5). The distribution of such remnants, when plotted on a modern map, does not make any sense at all. A vast ice sheet would be required to occupy most of the Indian and South Atlantic Oceans. Ice would have extended from south of the equator northwards to the Congo Basin and India. If however the continents are moved back into their original positions before continental drift took place, the ice age remnants combine to present a picture of a large land-based ice sheet, centred on the South Pole of the time (thought to have been situated in the vicinity of Zambia, southern Africa) (Fig. 5, 6).

Although proponents of a permanent fixed earth had no answer to Wegener's ice sheet argument and a less than adequate answer to the density question, Wegener in turn was completely overwhelmed by a question posed by geophysicists (who study the physical properties of the earth's surface and interior). They stated categorically that all their studies suggested that although the earth had a molten core, the remainder (i.e., all the earth's outer layers) behaved as a rigid mass. If this was so, how were the continents going to move? In other words, if Wegener's theory was tenable there should be a soft layer under the continents (in the form of molten or semi-molten rock) capable of flowing under pressure, and allowing the

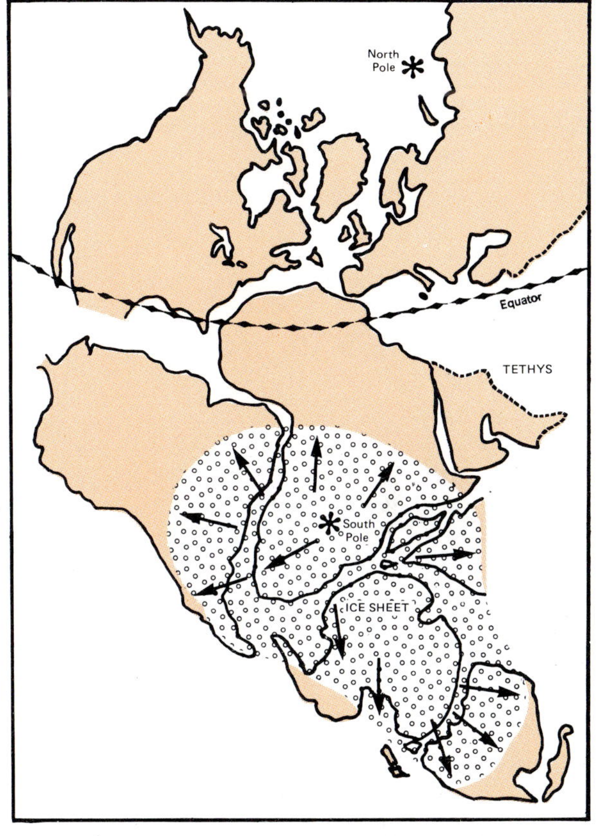

Fig. 6 The continents restored to positions it is thought they originally occupied in Carboniferous–Permian times. The continents that have remnants of the Carboniferous–Permian glaciation preserved on them were at that time close to the South Pole. Note that the directions of ice movement are consistent with a spread of ice outwards from the South Pole (compare with Fig. 5).

New Zealand was on the edge of the immense ice sheet, but at that time, little if any land was present in the New Zealand region and ice therefore did not accumulate. Countries like England, France, and the Netherlands were at that time very much closer to the equator than they are today, and experienced hot, arid conditions.

continents to move. Wegener initially published his theory in 1912 and revised and expanded versions appeared in the years between 1915 and 1924. Although he tried valiantly to answer his critics, his theory became a joke.

Scientific opinions began to change quite dramatically, however, in the 1950s and 1960s. Oceanic research firmly laid to rest the idea that the sea floor was littered with the remains of sunken continents and land bridges. Although some areas of continental rocks were found beneath the sea, it was shown that these comprised the worn-down edges of landmasses, covered by shallow seas (the so-called "continental shelves"), and the remnants of ancient lands worn down to such an extent that only isolated bits and pieces poke up above the sea. The Campbell Plateau is an example of a worn-down remnant of a formerly more extensive ancient New Zealand landmass. Apart from these exceptions, the deep oceans were found to be floored entirely with oceanic rocks. Areas of oceans can thus be classified as either true oceans ("oceanic basins") or marginal seas (continental shelves, submerged worn-down continental fragments, etc.).

CONTINENTAL JIG-SAW PUZZLES

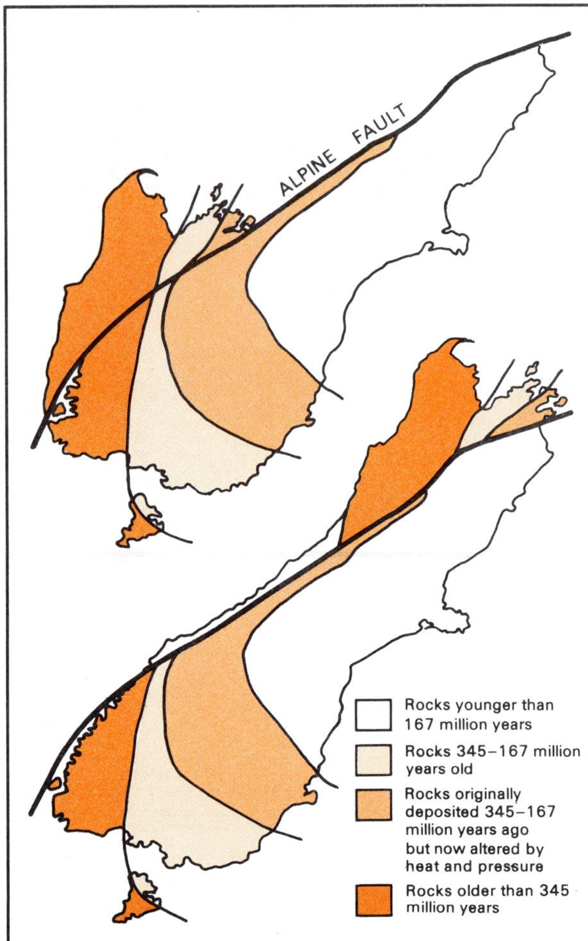

Fig. 7 Gradual movement along the Alpine Fault, spread over many millions of years, has torn New Zealand's South Island in two. Rocks that were once adjacent have moved for a distance of some 450 km. The upper diagram depicts the situation before movement took place (some people say about 140 million years ago, others 9 million years ago). The lower diagram depicts the situation as it is today.

In his time Wegener was asked to point to modern examples of continental drift. Critics of his theory said that if such continental movements had occurred in the past, then surely some signs of modern movement would be visible today. He was, however, unable to provide any examples.

Meanwhile, the progress of geological mapping has enabled geologists to gain a better picture of the geological structure of large areas of the world. In the 1940s and 1950s it was realised that in countries as widely separated as Scotland, California, and New Zealand, large blocks of land had moved considerable distances horizontally. Such movements had occurred along very large lines of weakness in the earth's crust, called faults. The existence of fault lines had been known from the early days of systematic underground coal mining in the 18th century. However, movements along most faults were thought to be primarily vertical (up or down), with horizontal (sideways) movements playing a very minor role (if any).

The matching-up of geology on either side of the Great Glen Fault in Scotland, the San Andreas Fault in California, and the Alpine Fault in New Zealand (Fig. 7), showed that large horizontal displacements had occurred. Such movements, measured in hundreds of kilometres, had considerably altered the geography of Scotland, California, and New Zealand. In short, this was continental drift in the making.

Somewhat later, in the 1960s, even larger horizontal faults were detected on the sea floor, displacing features over distances of several thousands of kilometres.

At about the same time significant developments were occurring in seismology (the study of earthquakes). These eventually provided a solution for the problem of a mechanism for continental drift.

Large earthquakes of magnitude 8 and above, although often terribly destructive in terms of human lives and property, are nonetheless of great scientific value. Such large earthquakes produce shock waves that bounce around many times inside the earth so that seismologists gain information that can be cross-checked and compared in great detail. Fortunately for humans, such large earthquakes do not occur very often. However, one originating in Kamchatka (eastern U.S.S.R.) in 1952 alerted seismologists to the possibility that the outer layers of the earth were not uniformly rigid, as hitherto assumed. These tentative observations were confirmed by other large earthquakes in Chile (1960) and Alaska (1964). Study of the seismic records obtained from these earthquakes showed quite clearly that the outer layers of the earth were divisible into rigid and non-rigid categories.

The outer skin or crust of the earth, comprising the continents and oceans, is composed of hard rigid rock (averaging 33 km in thickness under continents, 65 km under mountain ranges and 10–11 km under the seas). This layer is called the lithosphere (meaning "rock sphere"). The lithosphere is in turn underlain by a non-rigid layer of rock held at or near melting point, and probably having the consistency of hot toffee. This layer lies between 50 and 250 km below the surface of the earth and is called the asthenosphere ("weak sphere"). Underlying the asthenosphere is the mantle, a zone of rigid hard rock extending from a depth of 250 km to the outer boundary of the earth's core at 2900 km (Fig. 8).

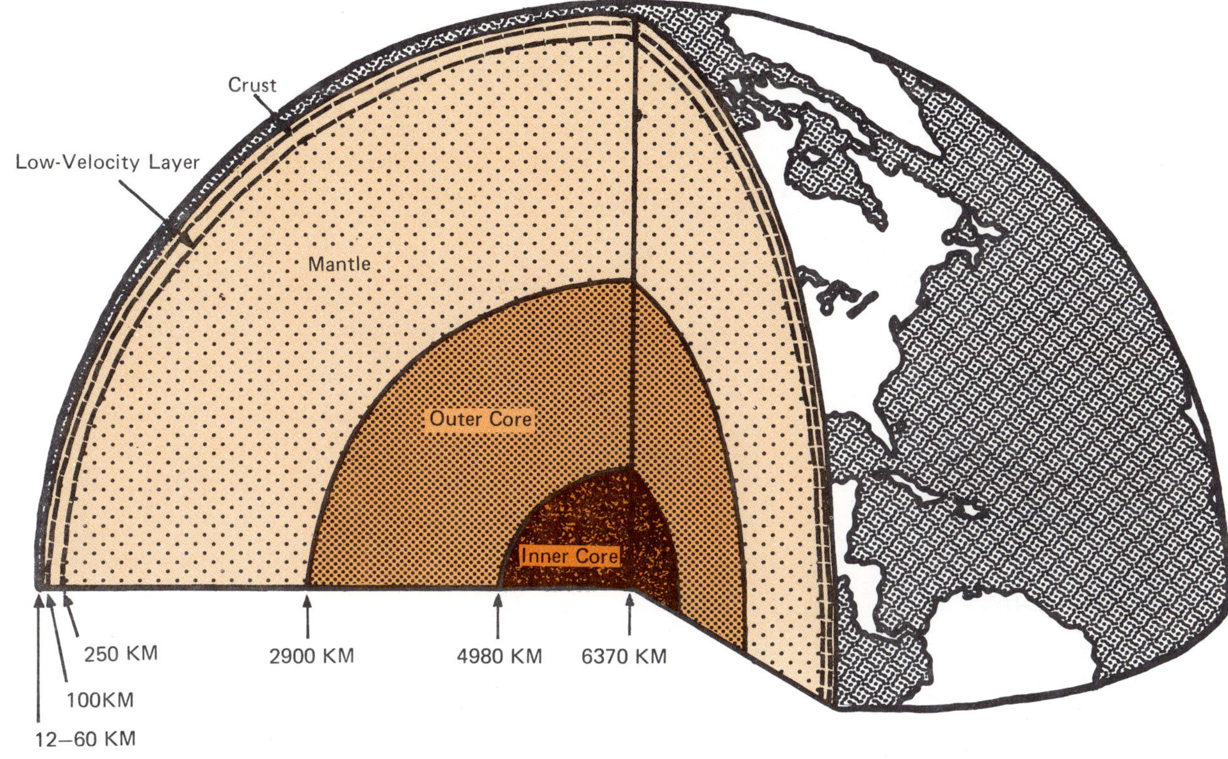

Fig. 8 Study of earthquake waves has shown that the earth is composed of a series of concentric layers of differing physical and chemical properties. The rocks of the thin crust or Lithosphere are cool and rigid. Mantle rock, which is hot, is capable of very slow movement. Mantle and crust are separated by a "low-velocity layer" or asthenosphere which has properties approaching that of a liquid. Evidence from earthquake waves indicates that the outer core is molten nickel iron, while the inner core is of the same material but solid. This diagram pictures the earth's layers as simple concentric shells and does not attempt to show the active processes that go.on in the interior.

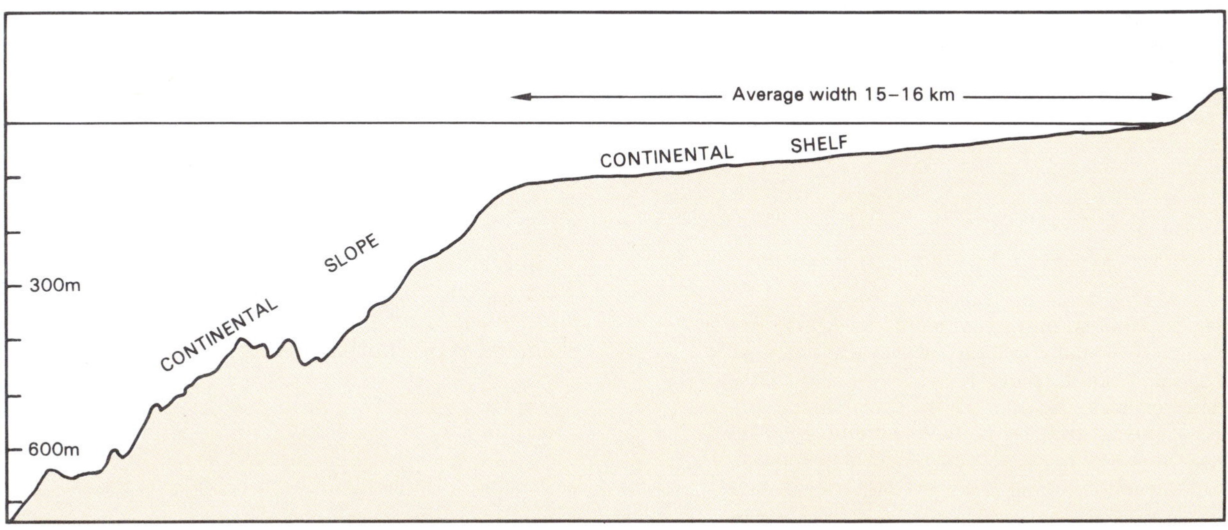

Fig. 9 Landmasses are surrounded by shallow or "continental shelf" waters and there is often a reasonably steep descent ("continental slope") down into deep abyssal water. Continental shelves around New Zealand vary greatly in character. They range from narrow fringes less than 8 km wide to extensive submarine plains, many tens of kilometres wide; from comparatively smooth platforms to areas of irregular relief. The example illustrated in this diagram is typical of the sea floor off the south Taranaki coast.

Fig. 10 Computer fit of the Atlantic continents. The fit was made at the edge of the continental shelf, using the 500-fathom (914 m) contour line of the continental slope. The edge of the continental shelf is dotted and the black areas mark gaps or overlaps. Although the individual pieces of the jig-saw puzzle have been apart for some 150 million years the overall fit is nonetheless very good. Many of the gaps may have been original inlets or lakes, between continental highlands, and some of the overlaps may result from a build-up of sediments in local areas during the 150 million years that have elapsed since separation. A notable example of an overlap with this latter origin is the Niger delta, responsible for a considerable overlap onto the northeastern corner of Brazil.

COMPUTERS TO THE RESCUE

The advent of the computer also helped refine ideas. In Wegener's time and in later years various attempts were made to assemble the continents using a jig-saw puzzle approach, however, this method was beset by problems. First, it was recognised that the true edges of the continent were **not** the modern coastlines, but rather the outer edges of the continental shelves and that this is where any jig-saw puzzle fit must be made. However, precise mapping of the continental shelves (Fig. 9) had to wait until the post-war years, when precision depth measuring instruments and computer-assisted plotting techniques were introduced. Second, it was realised that any fitting of coastlines had to be done on the actual shape of the earth (which, in fact, is more pear-shaped, rather than circular). Merely cutting up and fitting pieces of a map, with various distortions resulting from the projection of spherical features onto a flat sheet of paper, simply would not work.

By the late 1950s and early 1960s bathymetric mapping had accurately defined the outer edge of the continental shelf in many parts of the world. Also, computer-controlled mapping had come to the rescue of cartographers trying to grapple with the problems of spherical geometry. With the help of powerful computers, it became possible to compare the edges of continental shelves with a very high degree of precision. The results achieved were most impressive. In the South Atlantic Ocean, for example, the fit of the continents today is 99% perfect, notwithstanding the fact that the matching coastlines have been apart for some 90–130 million years (Fig. 10).

FOSSIL COMPASSES

The spinning earth acts as a gigantic dynamo, generating a strongly directional magnetic field which causes compass needles anywhere in the world to point towards the North Magnetic Pole. This directional effect is called polarity, and will be considered later. The other major effect of the earth's magnetic field is that the magnetic forces act on a compass needle in various ways depending on the latitude. Close to the equator the compass needle lies essentially flat, but closer to the poles the angle of the needle steadily increases, until near the poles the needle is virtually pointing straight down (Fig. 11). For this reason a magnetic compass is virtually useless in polar areas because the needle jams and cannot swing freely. Satellite navigation or celestial navigation using positions of the sun and stars are used. The angle (or declination) of terrestrial magnetism varies precisely with latitude and may be measured with great accuracy.

Remnants of past magnetism are preserved within many rocks, especially those rich in magnetic minerals. Iron occurs in most rocks. The iron mineral magnetite gets its name from its property of displaying natural magnetism. The first compasses used in ancient times by Chinese and Elizabethan voyagers used pieces of lodestone, a rock rich in magnetite. Magnetite is very abundant in igneous and volcanic rocks. However, active volcanoes often shower large areas of the surrounding country with volcanic ash, so many sedimentary rocks may also contain magnetite derived from the easily eroded ash. Erosion of igneous and volcanic rocks adds magnetite grains to a wide variety of other rocks. For some time it has been known that the magnetic properties of objects fade and eventually

disappear altogether with heating. In magnetite, for example, all magnetic properties are lost above 575°C (called the Curie Point). This observation led to the idea that when rocks are molten they are non-magnetic, as their temperatures are above the Curie Point. Thus lava is non-magnetic when it is poured out from a volcano, because its temperature is usually above 1000°C. The tiny grains of magnetite within the molten lava are randomly orientated, yet as the lava cools, and its temperature nears the Curie Point, the grains become magnetic and align themselves in response to the earth's magnetic field. They therefore act as a multitude of tiny compass needles, and as the rock solidifies, the direction and inclination of the earth's magnetic field at that time and place are, in effect, frozen into the rock.

Providing they have not been re-heated in more recent times, rocks such as ancient lava flows may be sampled today and measured for the magnetic direction and inclination of the time. For example, 180 million years ago large areas of western India were covered by vast sheets of lava. Since then, rivers have cut deep gorges through the lavas, exposing the many layers, like layers in a cake. In the early 1950s the exposed layers of lava were sampled and their magnetism studied. These studies showed that the angle of magnetism recorded in the ancient lavas corresponded to that of latitude 64°S, whereas today India lies between 7° and 35°N latitude. Comparable studies were made in several countries and all told a similar story: there was often considerable discrepancy between the latitude position indicated by the tiny compass needles in the rock and the modern position of the landmass.

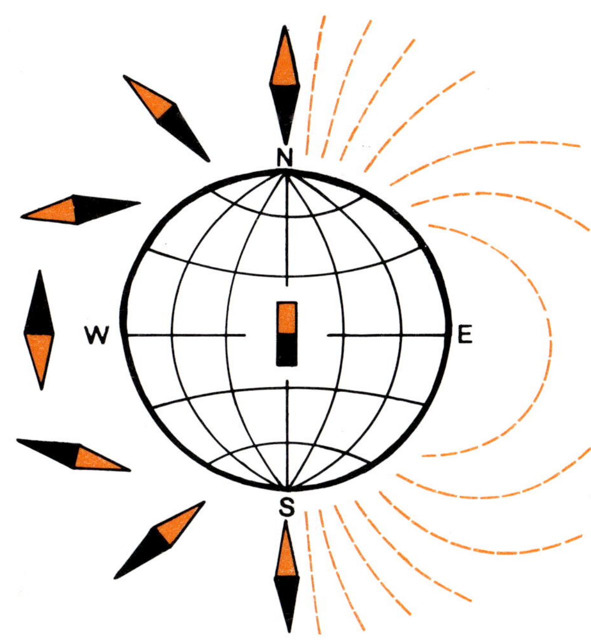

Fig. 11 A suspended magnetised needle, free to pivot vertically, will align itself in the direction of the earth's magnetic field. The inclination (or dip) of the needle varies from vertical over the pole to horizontal at the equator. Measurements of the angle of inclination of the magnetisation in rocks can therefore be used to determine the latitude in which the rocks originally gained their magnetisation.

At first it was thought that the earth's magnetic field had wobbled about in the past, but this changed to the idea that the magnetic field had remained relatively stable while the continents themselves had moved! Science was thus provided with a means of charting the past positions of continents with a fair degree of precision. Early studies in paleomagnetism (as it came to be called), begun in the 1950s, yielded the first quantitative evidence that continents had moved in the past. With data from other fields, they paved the way for general acceptance of continental drift.

SECRETS FROM THE SEAS

Submarine operations in both the 1939–45 World War and immediate post-war years demonstrated the need for accurate maps of the sea floor. A major effort was put into bathymetric mapping programmes. Areas of high natural magnetism offered opportunities for submarines to avoid detection by magnetic sensors, so it became increasingly important to compile magnetic maps of the sea floor. Spectacular progress was achieved in compiling bathymetric and magnetic maps of the sea floor. It was shown that the world's oceans concealed gigantic globe-encircling systems of submerged mountain chains, called mid-oceanic ridges. These ridges are associated with strings of submarine volcanoes and very high flows of heat coming from within the earth. The crests of many ridges were also found to be cleft by huge rift valleys, comparable to the Rift Valleys of East Africa. The submarine rifts were often found to measure 12–50 km across and to be flanked by steep walls 600–2000 m high.

Other areas of the oceans were found to be furrowed by enormous V-shaped trenches (Fig. 12), up to 3200 km long and 10 km wide. These trenches dwarf the mighty Grand Canyon, scooped out by the Colorado River in the western U.S.A., which by comparison is 320 km long and 1.6 km deep.

Chains of volcanoes, often in the form of island arcs, are closely associated with the trenches. Unlike the more tranquil basalt volcanoes along the ocean ridges, those of the island arcs are composed of a volcanic rock (andesite) that frequently erupts with great violence.

By the late 1950s and early 1960s mapping of the magnetism of the sea floor, and declassification of relevant naval information, had enabled oceanographers to discern patterns in the vast amounts of accumulated data. In particular they noted patterns of variation in the polarity of the magnetic "signatures" frozen into the rocks on the sea floor. "Polarity" is the property of the earth's magnetic field that ensures a compass needle always points towards the North Magnetic Pole. When analysed, the magnetic data showed certain areas of the sea floor to be composed of normally magnetised basalt rocks (with a magnetism orientated towards the North Pole). Other areas were reversely magnetised (with magnetism orientated towards the South Pole). It was evident that in the past the earth's magnetic field sometimes had a north-orientation whereas at other times it had flipped to a south-orientation.

Initially such polarity reversals were known only from the rocks of the deep sea floor—rather inaccessible to the geologist's

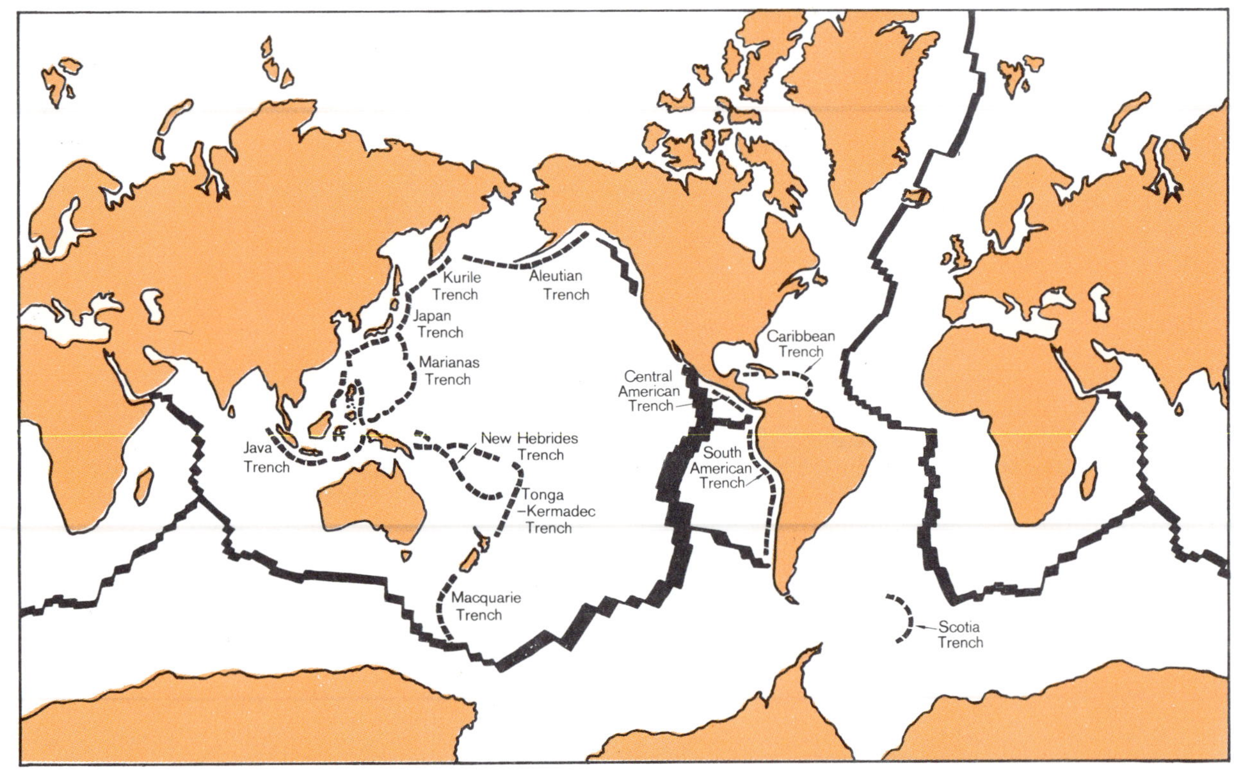

Fig. 12 Oceanographic research has revealed a globe-encircling system of oceanic mountain chains (oceanic "ridges" and "rises"), often disrupted at intervals by transverse fractures. These oceanic ridges and rises, where new oceanic crust is being created, are shown on this map as broad belts in solid black. Submarine trenches, marking the regions where oceanic crust is being pushed or drawn down into the mantle (along "subduction zones") are shown as broken bold lines.

hammer! Because there is really no substitute for a piece of rock in the hand (that can then be examined and analysed) it became imperative to examine changes in magnetic polarity of some land-based sequences of lava flows. Studies were made in Hawaii, Iceland, and elsewhere, and some of the flips in magnetic polarity were identified. Sequences of lava flows were found in which changes of magnetism were preserved. By dating the lava flows, a time scale was defined and applied in turn to the sea floor rocks. It was also found that many areas of the sea floor were "striped" with areas of normal and reverse magnetism and that such patterns of stripes were distributed symmetrically to either side of the mid-oceanic ridges.

The idea was proposed that the basalt rocks of the sea floor formed progressively at the mid-oceanic ridges. In these areas the crust is visualised as being continually torn apart along the ridge crests, with the resulting tears represented by rift valleys. The splits in the crust do not remain open for very long, but are promptly filled by hot basalt lava welling up from below. As the lava cools and solidifies its magnetic minerals take up the then-existing polarity of the earth's magnetic field. Pulling apart of the crust continues, however, and the freshly solidified lava is split down the middle and the two halves are again separated by a rift. Fresh molten lava then wells up into the new rift and on cooling is in turn imprinted with a magnetic signature. If the process

continues over many millions of years, entire areas of new sea floor will be created, carpeted with rocks bearing unique magnetic signatures, rather like a tape recorder. The whole process is termed "sea-floor spreading" and the constant creation of new sea floor is likened to a moving conveyor belt, with the continents being carried along as passengers.

The identification and mapping of characteristic magnetic signatures in sea-floor rocks, and their matching with dated polarity changes in sequences of lava flows on land, enabled a sense of time to be obtained. This gave earth scientists the means of calculating rates of sea floor spreading. Such calculations led in turn to

estimates of ages of the rocks flooring the world's oceans. As a result it became clear that most of the oceanic rocks were very young, compared with those of the continents, and have been created by sea-floor spreading extending over the last 130 million years (Fig. 13).

Such inexorable expansion of the sea floor cannot go on for ever, otherwise our seas and our planet itself would get larger and larger. So where does the expanding sea floor go? The deep-sea trenches of the world are thought to hold the key to this puzzle. They are regarded as areas in the crust where old sea floor is being actively consumed. The sea floor, riding on the conveyor belt of sea-floor spreading, arrives

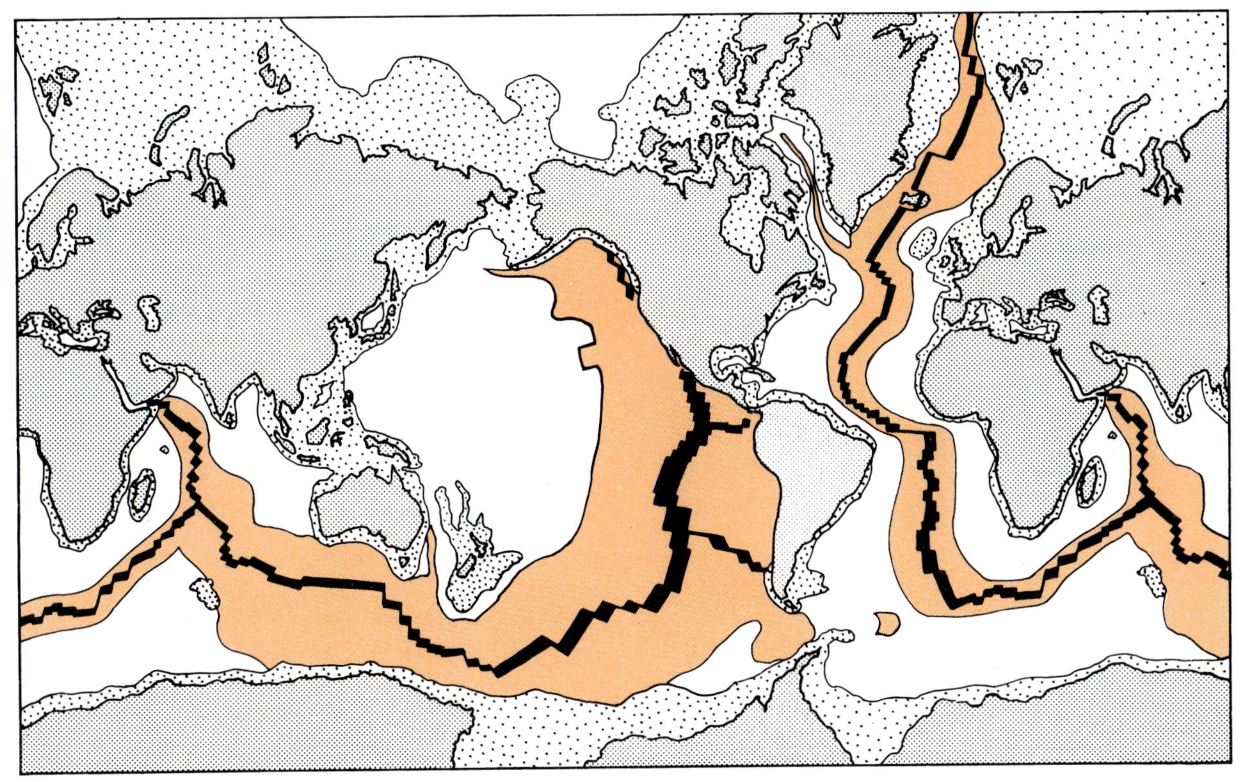

Fig. 13 Judging from magnetic studies and deep-sea drilling, vast areas of sea floor have been created in geologically "recent" times. This map shows the oceanic ridges and rises (in solid black), where new oceanic crust is being created. The coloured areas indicate the amount of ocean floor that has been formed over the last 65 million years. Submerged continental margins are shown by a coarse dot pattern.

at an oceanic trench where it is over-ridden and sinks obliquely deeper and deeper into the underlying mantle (Fig. 14). As it does so, swarms of earthquakes are generated by the enormous frictional forces involved in the downthrusting process. These earthquakes can be traced downwards into the earth as a sloping zone (the Benioff Zone) extending to depths of 600–700 km. The earthquake swarms of the Benioff Zone define the upper surface of the descending slab of sea floor. As the slab is pushed further and further down it enters into progressively hotter regions until it begins to melt. Also dragged down with the sea floor slab are the covering layers of sediment, much of it derived from adjacent continents and hence of continental composition. As melting begins, molten rock (magma) forms, containing a mixture of both sea floor and continental rocks. The magma finds its way to the surface, where it gives vent to particularly violent volcanic activity. Destructive eruptions occur at close intervals and massive cones are built-up that are in turn frequently torn asunder by awesome explosions. These volcanoes produce a distinctive volcanic rock called "andesite"— a name derived from the Andes of South America.

The general picture, therefore, is that the sea floor created at the mid-oceanic ridges is being swallowed up along "subduction zones"— marked by submarine trenches, andesite volcanism, and downward-dipping earthquake zones (Benioff zones) (Fig. 14).

The concept of sea-floor spreading, developed as a means of rationalising the magnetic patterns observed on the sea floor, remained a theory (albeit a rather elegant one) for a number of years. There was an urgent need to test these new ideas. If the oceans were really forming by a process of spreading out from mid-oceanic rises, there would be little if any sediment present on the sea floor close to the rises. Also, the thickness and age range of sediments should increase in directions away from rises.

Although some oil wells had been drilled in shallow coastal waters these had either used anchored barges or platforms fixed in place by deep piles. On the high seas, however, sampling of the sea floor was restricted to dredging and the dropping of weighted coring devices that sampled only the top-most layers on the sea floor. Clearly a major effort was required to increase knowledge of the oceans. As a consequence the U.S.A. in particular mounted a number of major oceanographic programmes. Research ships roamed the oceans plumbing the depths with sound waves and dragging magnetometers and dredges in their wakes. The most imaginative action, however, was to initiate the Deep Sea Drilling Project, using the drilling ship *Glomar Challenger*.

This ship, built specifically for the job, looks like a cross between an oil tanker and oil derrick. Bridge, living quarters, and six separate laboratories crammed with sophisticated apparatus are situated at the stern. Amidships, rising above an open hole passing right through the ship, is the derrick, its top towering 61 m above the sea. More than 7 km of drilling pipes are stacked on the deck, with more below in the holds. The ship can be held in position by means of twin propellors at both stern and bow, capable of being individually operated and under the control of a computer receiving signals from a beacon fixed to the sea bed. The effects of both winds and currents can thus be neutralised, allowing the ship to drill in some of the world's stormiest seas.

The main aim of the *Glomar Challenger's* programme was to obtain cores from the sediment layers in all the oceans of the world. Through their study, an overall picture of the age of the oceans would be built up and used to test the theory of sea floor spreading.

After sixteen years of operation and over 600 drillholes in all oceans and seas of the world except the Arctic, the Deep Sea Drilling Project has been an unqualified success. The project ranks as a technological triumph equal to exploration in space (and costing very much less!). It has been clearly demonstrated that the ages of the major ocean basins do vary in accordance with the predictions of the sea floor spreading hypothesis. A type of conveyor-belt mechanism, or something similar in effect, has been responsible for the opening up of the oceans.

Thus the results of new lines of research in such diverse fields as seismology, terrestrial magnetism, oceanography, computer technology, as well as the many "traditional" fields of geology such as petrology, paleontology, and stratigraphy paved the way for general acceptance of Wegener's ideas.

Fig. 14 Many earthquakes which occur under southern and central North Island have been found to originate in a reasonably well-defined zone, which slopes at 50° to the northwest. This crustal zone under New Zealand is a good example of a "Benioff Zone" and is interpreted as being the contact between the Indian–Australian Plate to the west and the Pacific Plate to the east.

Diagram (A), an east–west cross-section across the central North Island, shows the origins (epicentres) of earthquakes occurring within the shaded area of the key map. Active volcanoes are shown as triangle symbols.

Diagram (B) is an interpretation of the data presented in (A). The Indian–Australian Plate is being thrust over the Pacific Plate. As the Pacific Plate descends deep

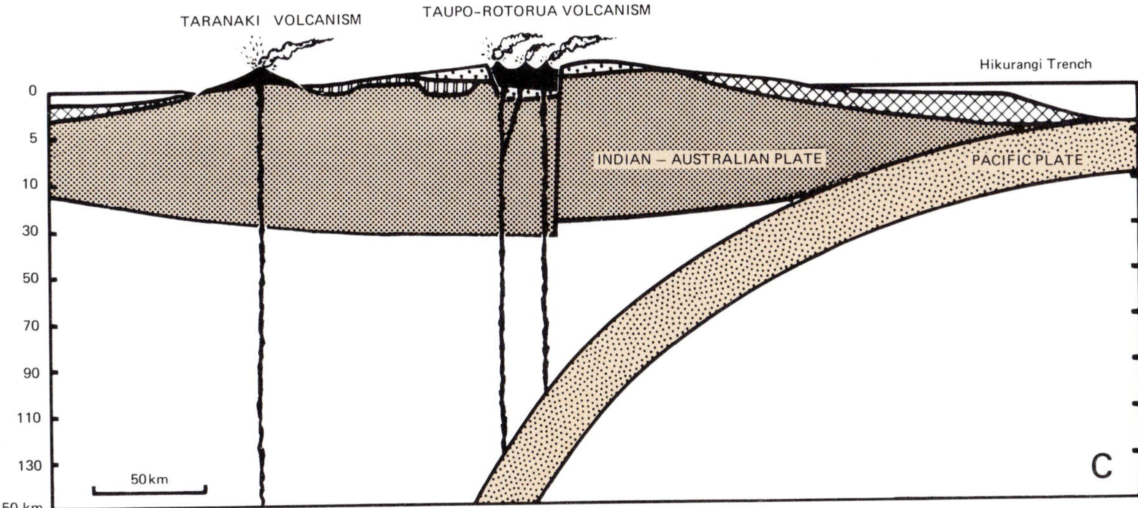

into the earth, numerous earthquakes are generated by the immense frictional forces, creating a "Benioff Zone". As the plate descends it encounters progressively higher temperatures, and melting occurs at various depths, corresponding to the melting points of the materials involved. The molten rock rises to the surface, breaking through the crust to form the active volcanic regions of Taupo–Rotorua and Taranaki. Deep earthquakes occurring at depths of 600 km under Taranaki, shown in (A), are thought to come from a piece of the subducted Pacific Plate that has become detached and has moved deeper into the mantle, as shown in (B).

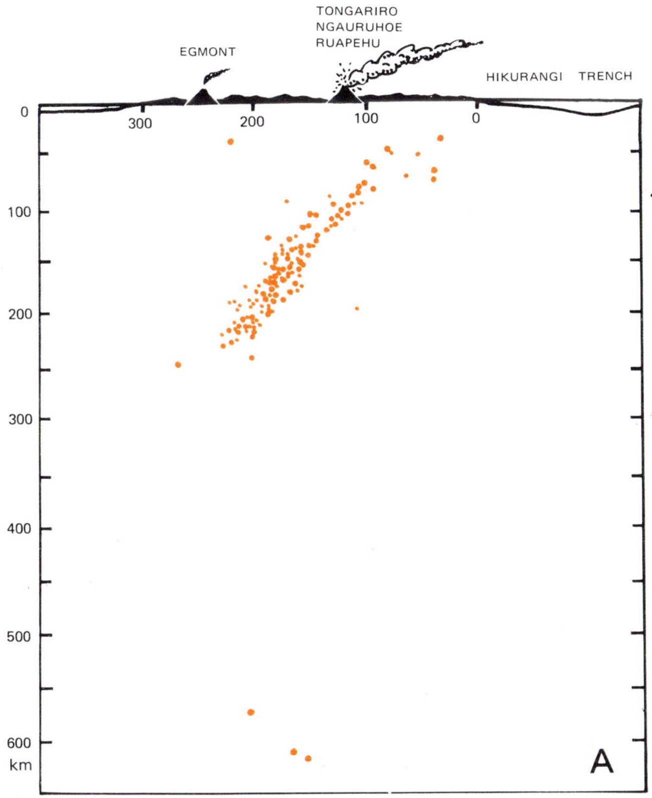

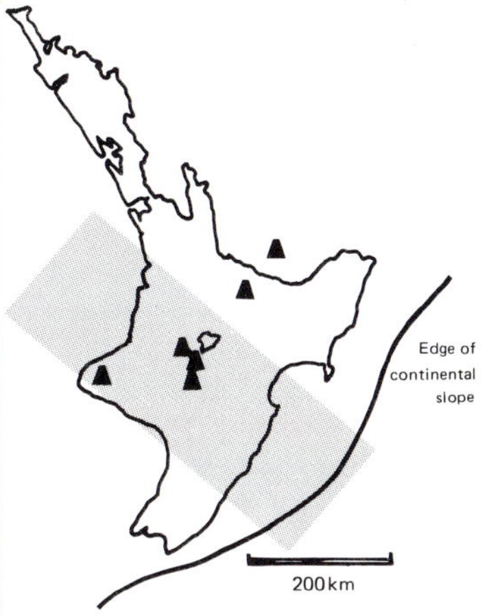

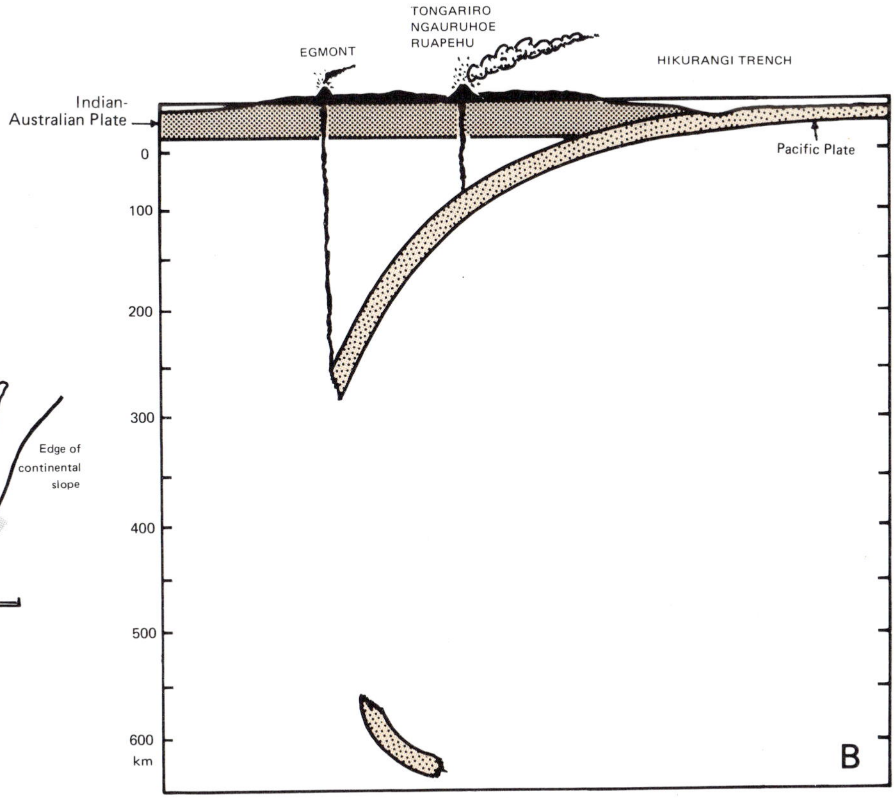

Diagram (C) is an enlargement of the upper part of (B), showing the relationship of the Taupo–Rotorua and Taranaki volcanic activity to the Benioff Zone.

MOVING PLATES

The new view of the ocean was not entirely compatible with Wegener's version of continents drifting around the world, floating, as it were, in a passive oceanic layer. Instead, the spreading of the oceans is regarded as the fundamental mechanism, with the continents being carried along like parcels on a conveyor belt.

At this point seismology again contributed to a major advance. The advent of seismometers of improved design and sensitivity, and their deployment world wide as a global seismological network enabled better fixes to be obtained on earthquakes occurring anywhere in the world. It became possible, with the aid of computers, to gain very precise maps showing the distribution of earthquakes. These maps showed that surprisingly large areas of the globe were essentially free of earthquakes. A large proportion of the world's earthquakes were found to be concentrated along very narrow zones. In turn, these were found to correspond with features such as mid-oceanic ridges, rift valleys, submarine trenches, island arcs (Fig. 12, 13) and giant fault lines such as those in California and New Zealand (Fig. 7). These observations gave rise to an entirely new concept, and a significant variation on Wegener's theme.

Instead of the world being viewed in terms of continents majestically moving around independently of each other, the earth's surface is seen as being made up of a patchwork of gigantic interlocking "plates". The plates are of varying sizes and all include continents as well as oceans (Fig. 15). Perhaps the most graphic way of visualising the situation is to compare the

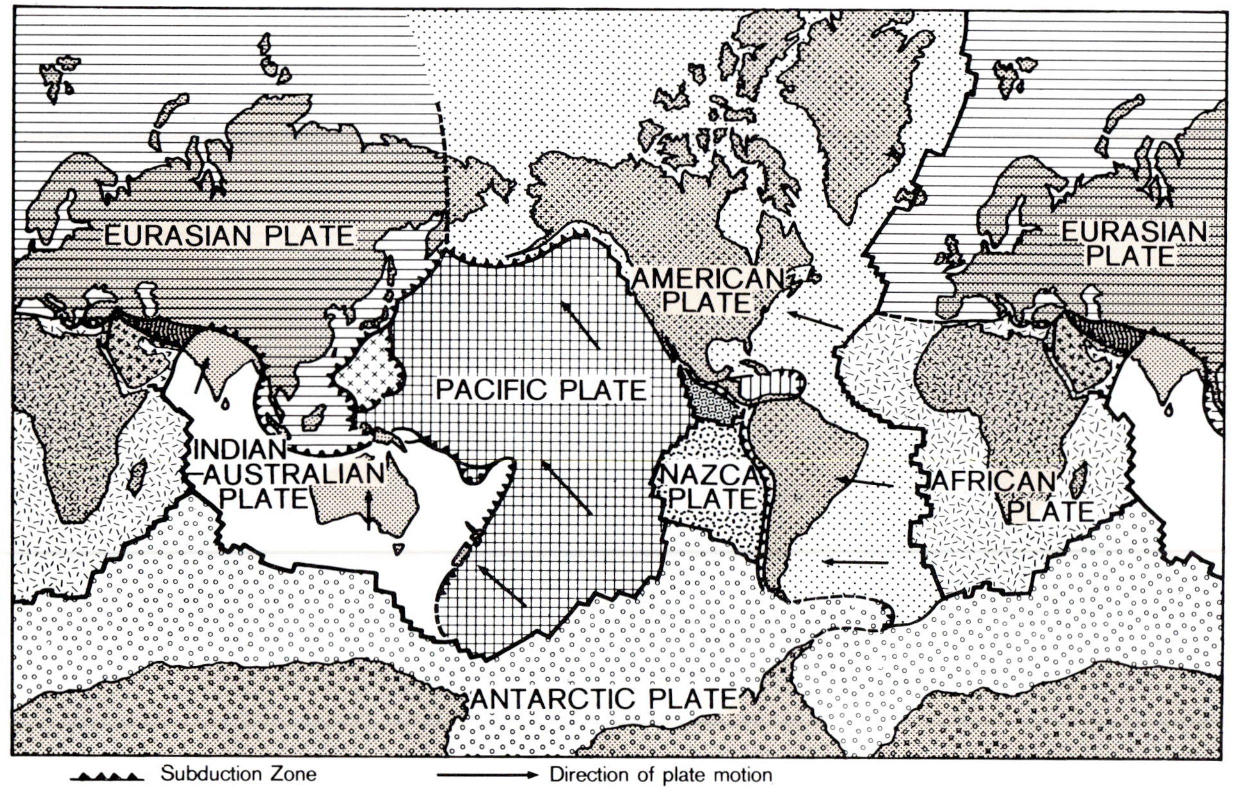

Fig. 15 The earth's crust is composed of at least 15 rigid virtually undistorted slabs or "plates" of lithosphere, 7 of which occupy considerable areas of the globe and are shown on this map. Boundaries of plates are of four principal types: divergent or spreading zones, where plates are separating and new plate material is being added; subduction zones, where plates converge and one plate is being consumed; "collision zones", former subduction zones where continents riding on plates are colliding; and transform faults, where two plates are simply gliding past one another, with no addition or destruction of plate material. Almost all the earthquake, volcanic, and mountain-building activity which marks the "active zones" of the earth's crust closely follows the boundaries of plates and is related to movements between them. As may be seen from the diagram, New Zealand straddles one such active zone, astride the boundary between the Indian–Australian plate and Pacific plate.

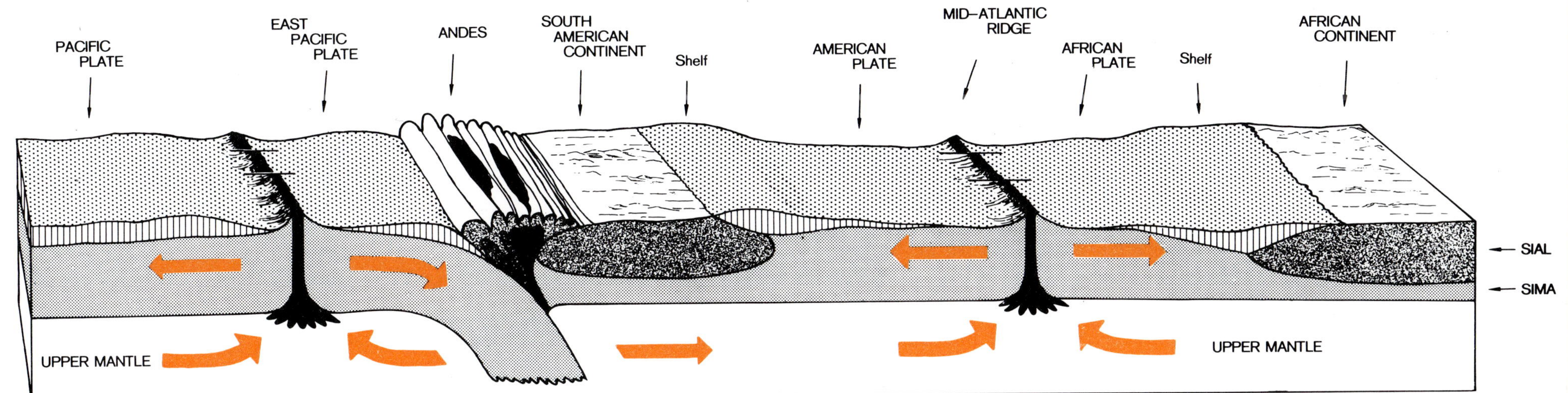

plates to ice floes on a frozen polar sea. As the ice floes move they jostle one another—sometimes colliding head-on and over-riding each other, sometimes scraping past sideways, or sometimes pulling apart from one another. However, regardless of the sense of motion, all of the "action" takes place along the edges of the ice floes, and their interiors remain unaffected.

So it is with the earth's plates: some edges are pulled apart and become the sites of mid-oceanic ridges, rift valleys, and active sea-floor spreading. Other edges mark the sites of collisions. If the collision is between two continents, the continents are welded together along a "seam" marked by rucked-up mountains. If the collision is between ocean floor and continent, or ocean floor and ocean floor, one side over-rides the other, an oceanic trench and island arc are formed, and sea floor material is actively consumed by melting in the mantle (Fig. 16). In other instances plates scrape sideways past one another with accompanying bruising (the rucking-up of mountains) and splintering (the formation of wide zones of faulted

terrain). The movements of plates can be tracked back in time and, when combined with information from fossil magnetism, deep sea drilling, and other geological research, a timetable can be established for the separation of the continents (Fig. 17).

The process by which various features of the earth's crust are created by movements of plates is called "plate tectonics". "Tectonics", meaning the building of geological features, comes from the same Greek root as "architect". Plate tectonics, therefore, is the building of global structures by means of interacting crustal plates.

Fig. 17   Information gained from deep-sea drilling and study of sea-floor magnetic patterns has enabled a timetable to be drawn up for the separation of the world's continents. This map shows the times, expressed in terms of millions of years from the present, at which the various continents began to separate from Gondwanaland.

Fig. 16   This is a stylised cross-section from the Eastern Pacific, across South America and the Atlantic Ocean, to Africa. It highlights the differences between regions where crustal plates are diverging (exemplified by the Mid-Atlantic Ridge and East Pacific Rise) and regions where plates are converging (exemplified by the western coastline of South America). Volcanic magma is welling up in the rifts along the Mid-Atlantic Ridge and East Pacific Rise, and also intruding into the Andes—the latter magma is a by-product of the melting of the descending plate.

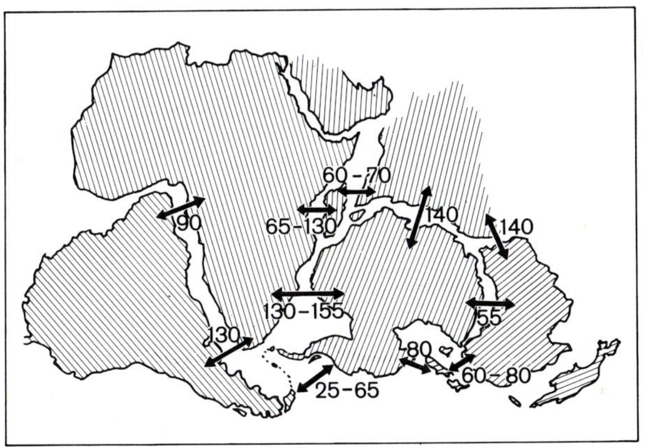

WHAT OF THE FUTURE?

Acceptance of the ideas of continental drift and plate tectonics has enabled us to look at the earth in a new light. We now know, for example, that the oceans are comparatively young, in geological terms (see Fig. 13). By contrast, large areas of the world's major continents are immensely old and the land itself has been built up as a complex mosaic of strips welded on at various times throughout geological history.

Projections based on calculated rates of sea floor spreading have allowed earth scientists to indulge in predictions for the future. However, plate movements are very slow in human terms. Greece, situated on one of the faster moving plates, has moved some 90–180 m relative to Africa since the time of Socrates (400 B.C.). Similarly, the Atlantic has opened up some 5–10 m since Columbus crossed it in A.D. 1492. The Atlantic is opening at a rate of 12–25 mm per year (12–25 km per million years). The Pacific is opening at a rate of 35–90 mm per year (35–90 km per million years). In terms of our own bodily growth, the Pacific is opening up at the rate one's fingernails grow (0.10 mm per day) and the Atlantic at the rate one toenails grow (0.05 mm per day). Thus as New Zealand (or at least some of it) drifts steadily northwards as part of the Indian–Australian plate, it will take 20 million years or more to feel the warmth of a tropical climate. Similarly, it will be 50 or 60 million years before Australia starts to nudge the Chinese mainland (Fig. 18).

As may be seen from the map of the world 50 million years into the future (Fig. 18), some major changes are envisaged that will considerably alter world climatic patterns. Continents will alter their latitudes and ocean currents rearrange themselves in the wake of such changes.

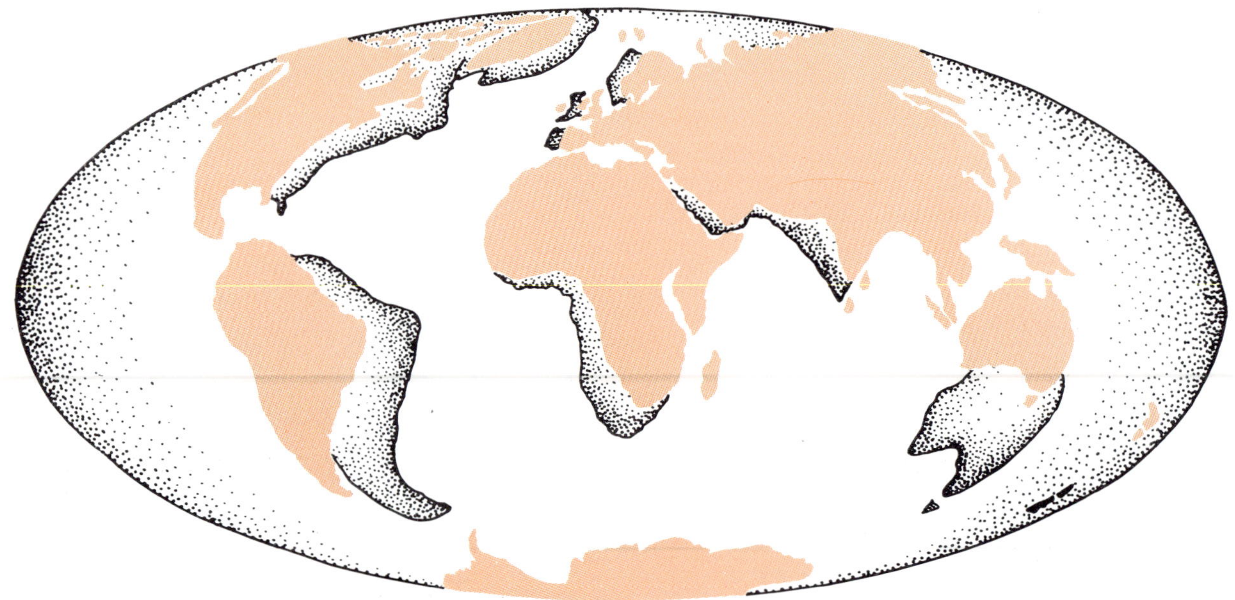

Fig. 18 Some 50 million years from now the world may look something like this. If in the future the world's plates continue their inexorable movements, the Atlantic and Indian Oceans will grow at the expense of the Pacific (unless the world expands simultaneously). Today's Red Sea has widened out into a mini-ocean and the East African Rift System is the site of an embryonic strait. The northern drift of Africa has squeezed Spain and Portugal into the Bay of Biscay and virtually closed off the Mediterranean. Australia and New Guinea have moved northwards to begin to nudge Eurasia, and New Zealand has similarly drifted in the same direction; although whether it will continue to be torn apart by the Alpine Fault, as shown in the diagram, or move as one piece on the edge of the Indian–Australian plate is not clear. The part of California west of the San Andreas Fault will be severed from the rest of North America and will begin drifting to the northwest. In about 10 million years Los Angeles will be opposite San Francisco, still fixed to the mainland. In about 60 million years Los Angeles will start to slide into the Aleutian Trench.

THE MAKING OF NEW ZEALAND

Adoption of the ideas of continental drift and plate tectonics has enabled geologists to take a new look at the earth. Fresh insights have been gained of the way modern New Zealand has been formed and of New Zealand's place in the geological past.

Although we know very little about how exactly the earth's plates were arranged in the past, we do know that New Zealand was probably part of Gondwanaland up to the time the Tasman Sea began to form, about 130 million years ago. For most of this time the site of modern New Zealand was an area on the ocean floor on the eastern edge of Gondwanaland, flanked by eastern Australia, Tasmania, and Antarctica (Fig. 3).

At times archipelagoes and scattered islands came into existence, only to be eroded away again. The shape of the New Zealand landmass as we know it today is a very recent feature and only came into existence some 10 000 years ago.

Although the present shape of New Zealand is shown on all the maps in this atlas, it should be remembered that this has been done for reference only, and that our land has had a great variety of shapes and sizes in the past.

As far as we can tell, from at least the Carboniferous onwards (from about 360 million years ago) the New Zealand area was the site of successively forming subduction zones and associated submarine trenches, troughs, and volcanic arcs. All were presumably related to a long history of convergence and interaction of crustal plates, and probably involved progressive underthrusting of the floor of the ancestral Pacific under the eastern edge of Gondwanaland (Fig. 19).

Active subduction and volcanic activity generated from the volcanic arc continued throughout the Permian, Triassic, and Jurassic (300–135 million years ago). Substantial thicknesses of sediment were laid down on either side of the volcanic arc. All this time more and more sea floor was being conveyed to and consumed in the subduction zone. Eventually a mass of sediment, originally deposited off the coast of West Antarctica, arrived at the subduction zone and, being buoyant continental material, could not be consumed in the subduction zone. It rode up and collided with the sediments that had accumulated in the submarine trench and on the flanks of the volcanic arc. Immense thrusting, shearing and buckling movements resulted and slices of sea floor and upper mantle were caught up and pushed into the contorted piles of sediment that were forced up to form land. These major earth movements, collectively called the Rangitata Orogeny or mountain-building phase (from the Greek *oros*, a mountain), are thought to have occurred mainly during the late Jurassic and early Cretaceous (140–120 million years ago) (Fig. 20). The ancestral New Zealand created in this way, was the first sizeable landmass to exist here. It extended northwards to at least New Caledonia and southwards to Campbell and Auckland islands.

The resulting accumulation of a contorted mass of continental material, too buoyant to be subducted down into the mantle, "gummed up" the subduction zone, causing the site of active subduction to migrate to the edge of the newly created land (Fig. 20).

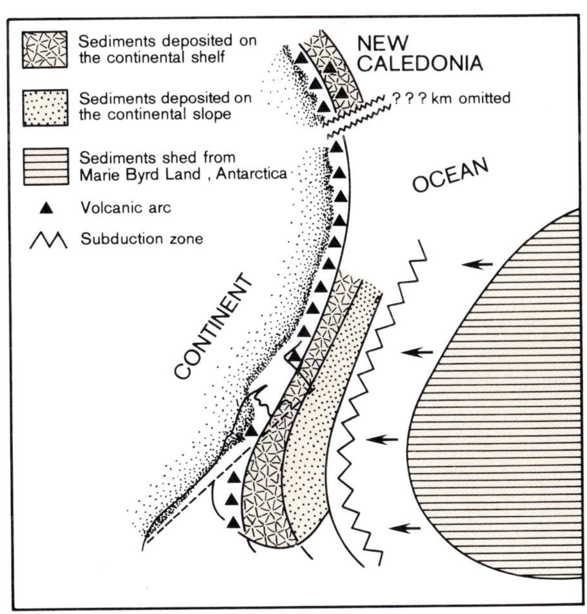

Fig. 19 Virtually identical rocks and fossils were laid down in New Caledonia and New Zealand during Permian, Triassic, and Jurassic time (280–136 million years ago). The two regions were part of a continuous trough of deposition (New Zealand Geosyncline), and were also close together geographically. The New Zealand Geosyncline bordered a large landmass (Australia and the Lord Howe Rise) and was in turn flanked by a subduction zone. Sediments shed from the Lord Howe–Australia landmass progressively filled the western flank of the New Zealand Geosyncline and belts of sediments laid down on the continental shelf and continental slope (Fig. 20) can be distinguished. Volcanic island arcs, related to the down-dipping subduction zone, formed extensive archipelagos in the New Zealand Geosyncline. As subduction continued, a mass of sediment derived from Marie Byrd Land was rafted across from Antarctica (cf. Fig. 20). The collision of these sediments with the westerly-derived material already deposited in the New Zealand Geosyncline generated massive earth movements (between 136–110 million years ago), that produced a primaeval New Zealand continent. This diagram is a sketch showing the geological relationships of New Caledonia and New Zealand just before such earth movements began.

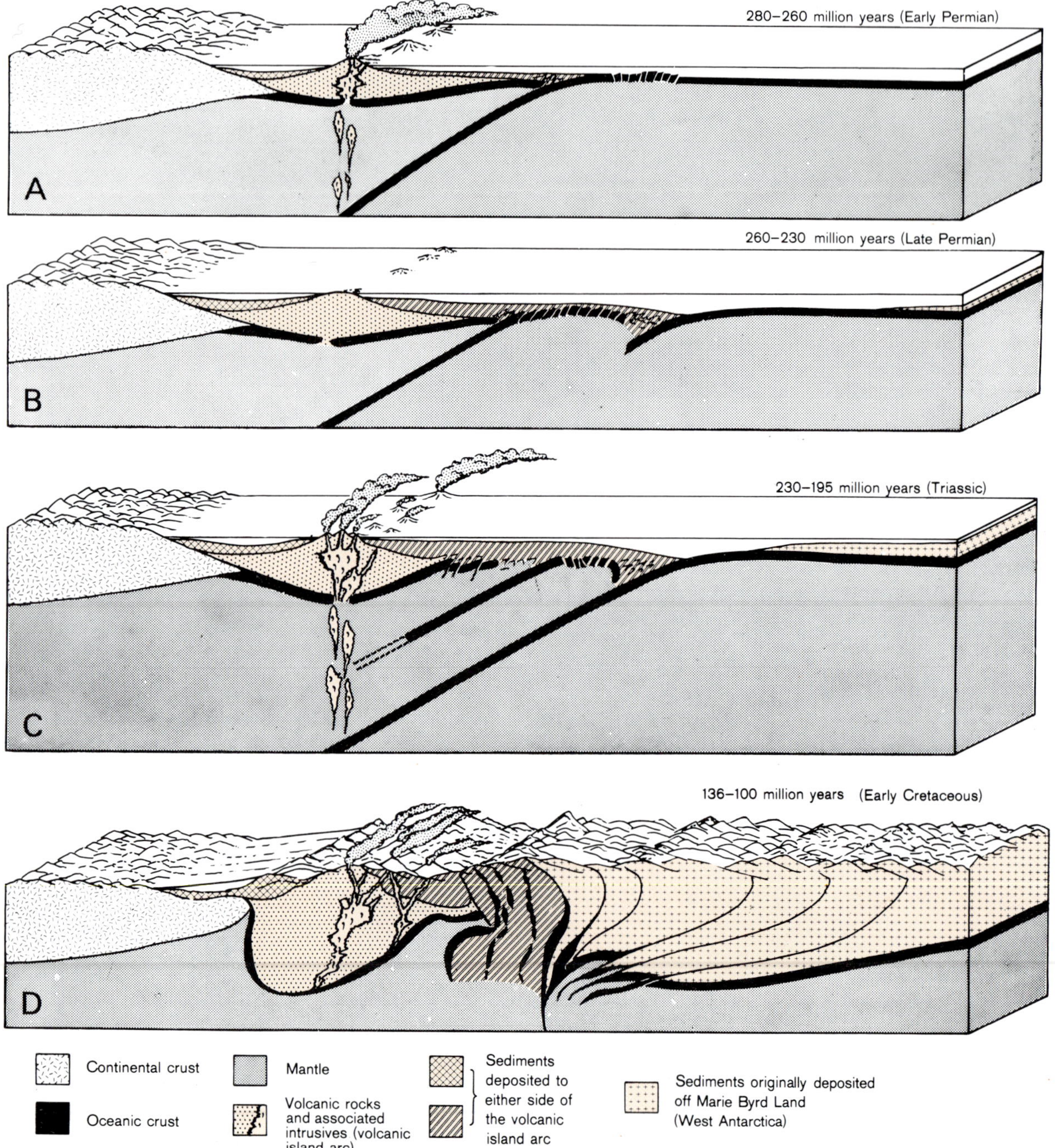

Fig. 20 Various models have been proposed to account for the geological features of the late Palaeozoic and Mesozoic of New Zealand and one is summarised in these diagrams. In the early Permian, some 280 million years ago (diagram A), the ancient continental rocks of eastern Antarctica, the Lord Howe Rise, Australia, and the West Coast of the South Island of New Zealand were being actively overthrust across the floor of the ancient Pacific Ocean, and a subduction zone was forming along the continental margin. Melting of the subducted slab generated an oceanic island arc, with active volcanoes. Sediment derived from the continents was laid down between the continental shoreline and the island arc, whilst volcanic-derived material accumulated to either side of the arc.

In the late Permian (diagram B) the subduction zone was disrupted by a change in plate movements and a new subduction zone formed to the east. Volcanic island arcs and thick sedimentary accumulations occupied the site of modern New Zealand, flanked to the west (left) by the ancient continental rocks of the Lord Howe Rise, Australia, and eastern Antarctica.

To the right (east) of diagrams B and C sediments laid on the sea floor off Marie Byrd Land (western Antarctica) are being steadily rafted towards the left (west), as ocean floor is steadily consumed in the subduction zone. These sediments are thought to be ancestral to the rocks that today largely form the alpine backbone of New Zealand. It is believed that throughout the Jurassic (195–136 million years ago) more and more sediments that had originated from Marie Byrd Land arrived at the subduction zone. This sediment, derived from continental material and therefore relatively light compared with oceanic crust and mantle, could not be consumed by the subduction zone, and so piled up and collided with the island arc and associated western-derived material. At this time (late Jurassic–early Cretaceous) the subduction zone died out and the two masses of sediment (from western and eastern sources) were folded, squeezed together, and buckled up to form an ancestral New Zealand landmass. Diagram D gives an interpretation of the scale and nature of the movements involved. In the late Cretaceous (100–80 million years ago), the ancient continental rocks of Lord Howe Rise and Australia split away, to leave New Zealand as a segment of newly-created continental rocks.

At about this time, sea-floor spreading began between ancestral New Zealand and Australia, and the embryonic Tasman Sea came into being (Fig. 21). The spreading ridge in the Tasman was active betwen 80 and 60 million years ago. During this time ancestral New Zealand was rafted eastwards and pushed further and further over the Pacific floor. The rock and sediments of the ocean floor progressively disappeared down the subduction zone lying along New Zealand's eastern margin.

The focus of active sea-floor spreading then shifted to between Australia and Antarctica. As a result these two lands that up to then had been in close contact, started to move apart, forming the Southern Ocean. These movements began about 55 million years ago, and are still continuing today. With the end of sea floor spreading in the Tasman Sea some 60 million years ago, New Zealand became fixed in relation to Australia. In other words, it became part of the Indian–Australian plate, and as that plate pulled away from the Antarctic plate, and drifted northwards, so did New Zealand.

Today New Zealand straddles the boundary between the Pacific and Indian–Australian plates (Fig. 22, 23). To the north of New Zealand the structure of the plate boundary is reasonably straight-forward. The Pacific oceanic floor is being subducted under the edge of the Indian–Australian plate. This is accompanied by the development of a submarine trench, Benioff Zone, volcanic island arc, etc. (the Tonga–Kermadec system) (Fig. 22, 23).

A similar plate boundary structure exists to the south of New Zealand (the Macquarie system). This boundary is complicated by the fact that here the edge of the Pacific plate is being pushed **over** the Indian–

Fig. 21 This reconstruction shows the Southwest Pacific at the time the Tasman Basin was beginning to open, some 80 million years ago. Also beginning to open at about the same time or later was the sector of the Southern Ocean lying between Antarctica and the edge of the Campbell Plateau. To arrive at the reconstruction, the sea-floor spreading process has been reversed, using ocean magnetic patterns as a guide, and the magnetics and bathymetry have been matched by computer. The geology of the various sub-Antarctic islands (Campbell, Auckland, Snares, Bounty, Antipodes, etc.), combined with information on sea-floor drilling, dredging, and magnetism, indicates that the geology of the South Island of New Zealand continues to the edge of the Campbell Plateau and the Chatham Rise and that the older rocks of both areas and those of alpine New Zealand came originally from the vicinity of Marie Byrd Land, West Antarctica.

This reconstruction shows the Chatham rise and Campbell Plateau set against Marie Byrd Land. The Bounty Trough (between the Chatham Rise and the Campbell Plateau) has been closed and so also has the Solander Trough, south of Fiordland. On the other side of New Zealand, the Tasman Basin and New Caledonia Basin have been partially closed. The New Zealand block has been regarded as a flexible unit and various distortions have been introduced. The parts of New Zealand on either side of the Alpine Fault (see Fig. 7) have been reversed back along the fault and rotated. A large shift of 1000 km is proposed on the Alpine Fault. The gap (shaped like a pie segment) between the two halves of New Zealand is thought to represent the amount of crust consumed in the New Zealand subduction zone during the past 80 million years. Although most geologists now accept that New Zealand, New Caledonia, Australia, and Antarctica probably fitted together something like this there is still a difference of opinion regarding the geometry of the fit, the amount of distortion undergone by New Zealand, and the amount of displacement and timing of movement along the Alpine Fault. The diagram indicates the fit along the 2000 m submarine contour. Landmasses are indicated by the darker colour, continental shelves by the lighter colour. Sea-floor spreading is pictured as actively taking place on either side of an oceanic ridge in the embryonic Tasman Sea.

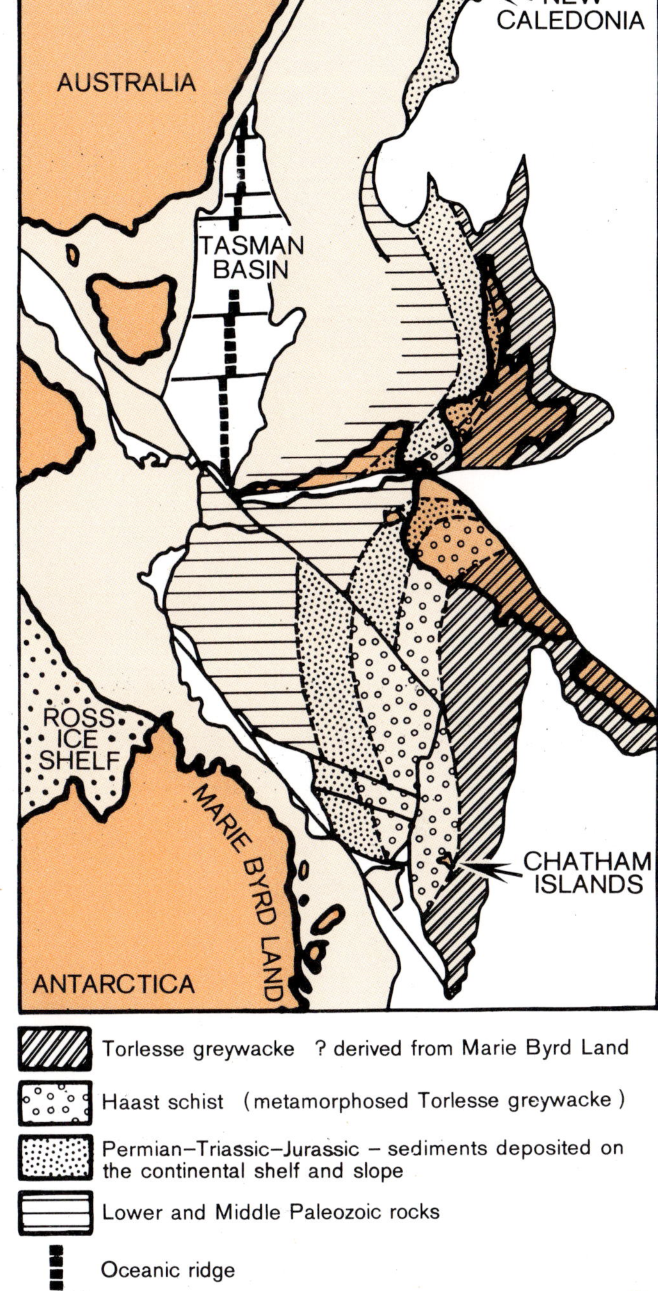

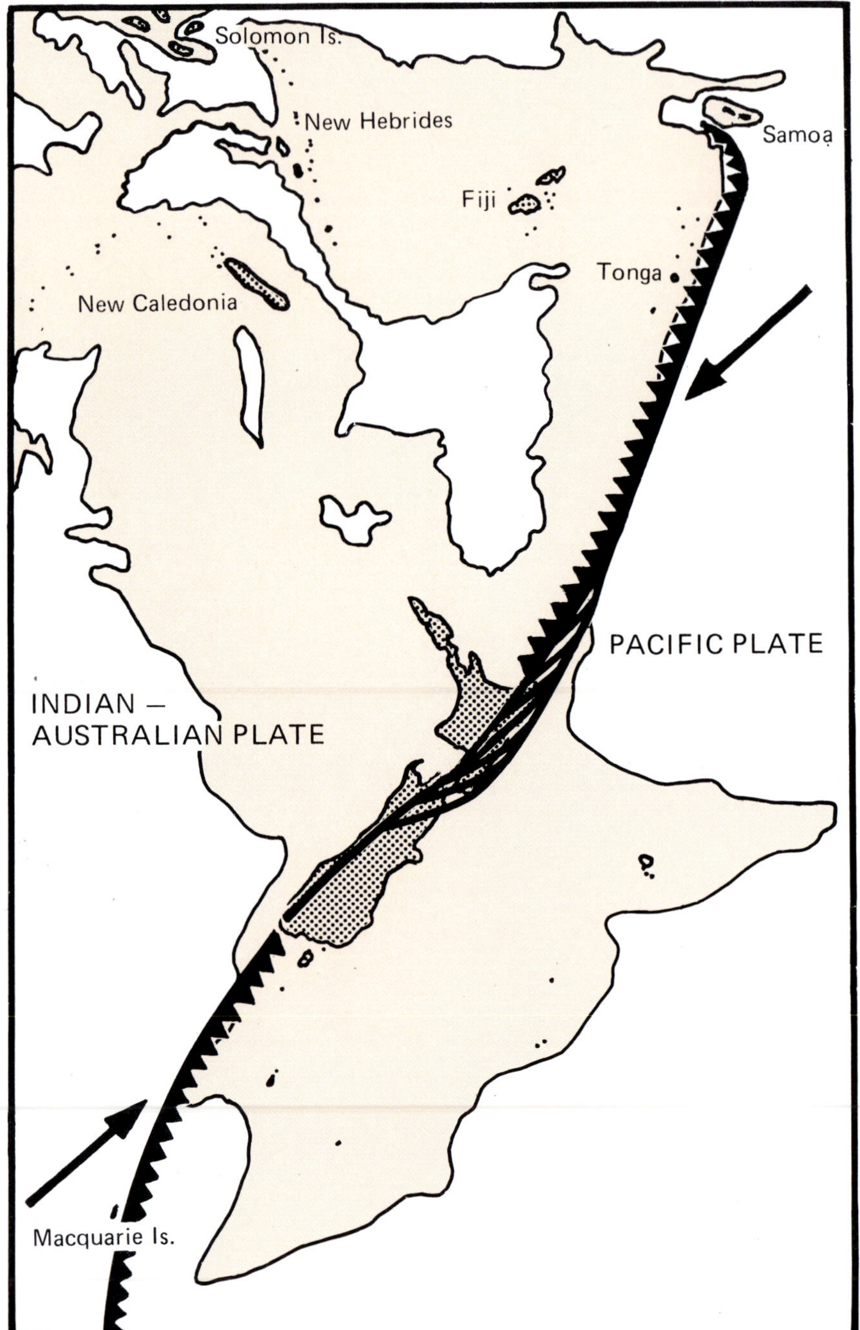

Australian plate. In this region all the features associated with subduction are developed as a mirror-image of the Tonga–Kermadec situation (Fig. 22).

New Zealand forms the link between the two, with the Tonga–Kermadec system dying out under the North Island, and the Macquarie system dying out as it approaches Fiordland.

As may be expected, the geology of New Zealand reflects the complexities of being the go-between for the two systems; having been squeezed, pushed, and pulled, and eventually torn apart. The Alpine Fault is one major feature that stands out amongst the maze of other related structures, and it, along with associated branch faults, forms a fault system, linking the two subduction systems (Fig. 22). Throughout its long geological history New Zealand has played a variety of roles, but has never been very far away from an active plate margin.

The future of New Zealand, geologically speaking, is difficult to foretell. If the boundary between the Indian–Australian plate and the Pacific plate remains in its present position, virtually bisecting the country (Fig. 22, 23), and if the Indian–Australian plate continues to move away from the Antarctic plate (a process that started about 55 million years ago), New Zealand may at some stage in the future find itself as two groups of islands as shown

Fig. 22 New Zealand's Alpine Fault (Fig. 7) links the Tonga–Kermadec Trench with the Macquarie Trench. In the North Island and the northern part of the South Island the Alpine Fault splits and forms a complex zone of individual faults (the New Zealand Shear Belt), that together pass into the Tonga–Kermadec Trench. The arrows show the directions of relative motion between the Pacific and Indian–Australian plates and lie on the plate edges being carried down into the trenches (depicted by the toothed pattern). The trench system extending through the New Hebrides and Solomon Islands and linkage via Fiji is not shown. Sea floor ranging in depth from 0 to 4000 m is indicated by the coloured area.

in Fig. 18. In this scenario the part of New Zealand west of the Alpine Fault is carried northwards as part of the Indian–Australian plate, leaving the part of New Zealand east of the Alpine Fault sitting on the edge of the Pacific plate. The people currently seeking independence for the South Island may therefore have their way eventually, although they will have to reconcile themselves to the loss of the West Coast and the addition of the Wairarapa and Hawke's Bay! Pieces of Wellington City may also go in opposite directions, so we may end up with a divided capital!

An alternative scenario views the present New Zealand geological situation as being transitional. In this view the situation has been complicated by the arrival at the plate edge of that part of New Zealand now lying east of the Alpine Fault (Fig. 7). Because it is continental (and therefore buoyant), this segment of land will probably not go down the Hikurangi–Kermadec-Tonga Trench. If this is so, the scene may be set for movement of the plate boundary eastwards, out into the Pacific, as happened in the Permian and Triassic (Fig. 20). If this happens, all of the New Zealand landmass will drift steadily northwards, as part of the Indian–Australian plate and some 20 million years into the future New Zealanders will be basking in tropical sunshine!

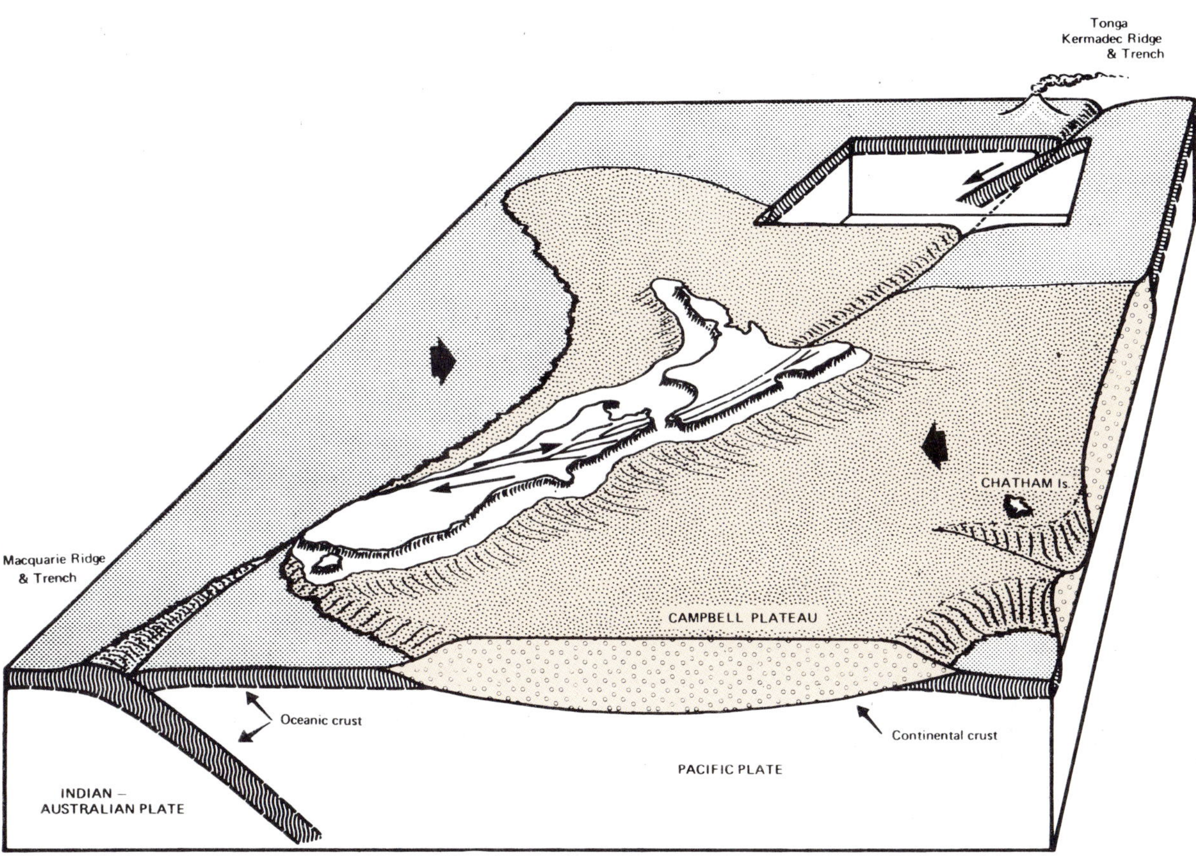

Fig. 23 Birds-eye view of New Zealand and its areas of submerged continental margin (the coloured areas). Together they represent a piece of continental crust straddling the boundary between the Indian–Australian Plate and the Pacific Plate (compare Fig. 15). The general sense of movement between the two plates is shown on the diagram by the large broad arrows. Southwest of New Zealand the Pacific Plate is being pushed over the Indian–Australian Plate (as shown in cross-section, left foreground). The opposite is happening northeast of New Zealand (as shown in the cut-away box, top right). Two opposite-acting subduction systems have developed: the Tonga–Kermadec system and the Macquarie system.

The New Zealand landmass and its submerged continental shelves, caught in a scissor-like position between the subduction zones, has been twisted and distorted and torn by huge fractures (faults). The major faults traversing the New Zealand are shown diagrammatically. Arrows indicate the major sense of horizontal movement along these faults, although such movements are usually also accompanied by vertical movements. Over geological time the vertical movements have created much of New Zealand's mountain scenery.

RECONSTRUCTING THE GEOGRAPHY OF THE PAST

By using the full range of techniques and accumulated knowledge summarised in the preceding pages, the past movements of the continents can be tracked with a fair degree of precision. We are therefore able to reconstruct a series of maps showing the geography of past times. Such a series is presented as Fig. 24–26. The maps begin in the Cambrian period of geological time, 570–500 million years ago. At that time, Australia and New Zealand were part of the super-continent Gondwanaland, made up of the present-day Southern Hemisphere continents plus India. The remaining continents were grouped, sometimes only rather loosely, as the super-continent of Laurasia. In the Cambrian, Gondwanaland was orientated in such a way that Australia and New Zealand lay in the lower and middle latitudes of the Northern Hemisphere and were close to Asiatic U.S.S.R., China, and Indo-China, with whom they shared a variety of marine creatures. In succeeding periods of time, a slow rotation of Gondwanaland progressively swung Australia and New Zealand southwards so that by Middle Silurian times (420 million years ago) New Zealand was positioned at about 30°S latitude. This southerly drift continued in the Devonian period, (410–350 million years ago), when contact was lost with China and Southeast Asia. In the Permian (300–235 million years ago) Australia and New Zealand were close to the South Pole and felt the effects of continent-wide glaciation.

In the succeeding Triassic and Jurassic periods (235–135 million years ago) a continuation of the same drifting movements progressively rotated eastern Gondwanaland away from the South Pole. New Zealand and Australia were slowly returned to middle latitudes. This trend was short-lived, because splitting movements along the future sites of the Indian and South Atlantic oceans separated eastern from western Gondwanaland and again swung New Zealand and Australia into positions close to the South Pole. This time, however, unlike the situation in the Carboniferous and Permian, large land masses were absent in the polar areas, and as a result extensive continental ice-sheet glaciation did not develop. (Large ice sheets can only accumulate on reasonably extensive and elevated areas of land.) Although in the early and middle Cretaceous (135–95 million years ago) Australia and New Zealand were in southern latitudes that today experience frigid sub-Antarctic weather, because of the absence of polar ice during these times, they instead experienced cool or cold-temperate climatic conditions.

New splitting movements, heralding formation of the Tasman Sea (between 80–60 million years ago) and the sector of the Southern Ocean between Australia and Antarctica (from 55 million years onwards), gradually moved Australia and New Zealand northwards away from the South Pole, towards their modern geographic positions.

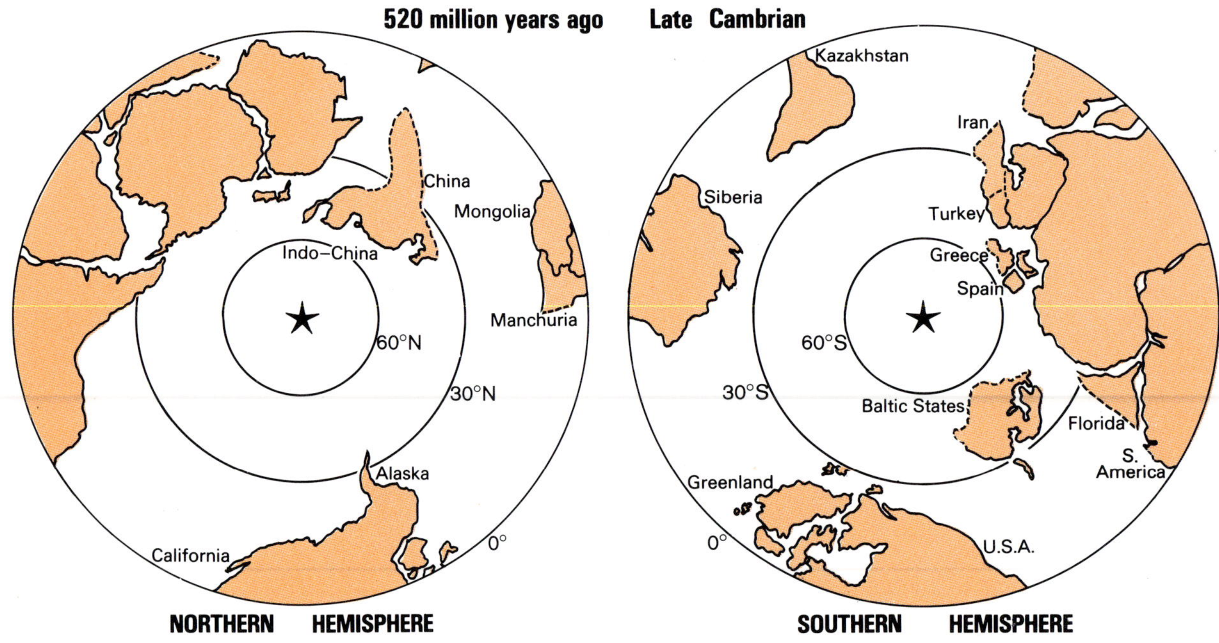

NEW ZEALAND'S CHANGING GEOGRAPHY

As may be expected, the gyrations of New Zealand around the world throughout the geological ages had major effects on its past fauna and flora, depending on the climatic zone in which New Zealand was situated at various times.

Climate, however, is only one of the factors influencing the character of New Zealand's past fauna and flora. Timing of the events of grouping together and splitting apart of individual continents is another rather obvious factor. Equally crucial is the nature of the links between continents: that is, whether suitable migration routes were available at particular times to permit movements of animals and plants. If continents were close together and had continuous land links and similar climates the chances are they were able to share the same land flora and fauna. If they were separated by a stretch of shallow sea, even of no great width, this would present a major obstacle to land organisms, while still permitting interchange of shallow-water marine faunas and floras.

In general, New Zealand has always been accessible to marine organisms with good means of dispersal. These organisms include forms with larvae having the ability to swim or float in the upper levels of the sea for extended periods, and/or adults with the same ability. Depending on factors such as wind and sea currents and temperature, such organisms were able to spread virtually world-wide. It is these groups that were free to migrate to New Zealand virtually at will throughout geological time.

Other organisms lacking such means of dispersal are entirely dependent on provision of suitable migration routes. They are very vulnerable to changes in factors such as geography, climate, wind and current conditions. Land organisms are obviously prime candidates for such restrictions. Most land animals have only very limited swimming abilities and for the majority a stretch of open sea constitutes a major barrier to movement. However, chance migration may sometimes occur— as instanced by the rafting across the sea of animals (and plants) on floating masses of vegetation torn away from the land by storms. Animals with flying ability are in a rather different category. Although they may

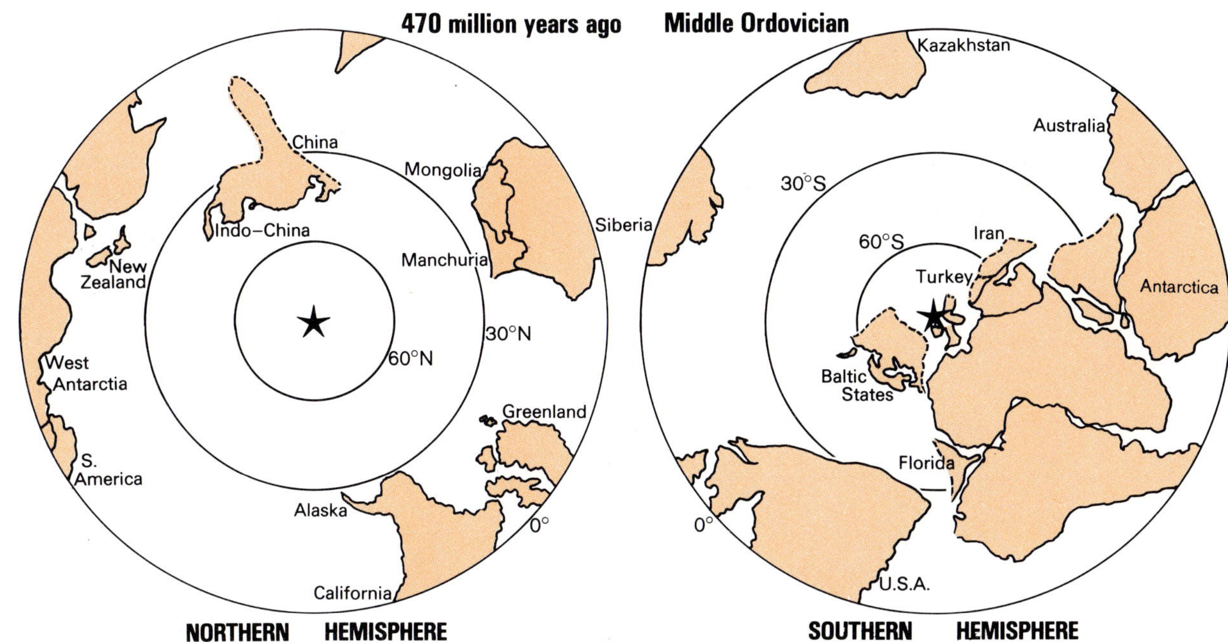

Fig. 24, 25 A world geography vastly different to that of today was evident in the Cambrian and Ordovician periods 570–435 million years ago. Studies of fossil magnetism preserved in rocks laid down at the time indicate that New Zealand and Australia were then in northern latitudes, close to modern Kazakhstan, China, and Indo-China. Later, however, global movements progressively rotated New Zealand, Australia, and other Gondwana countries into the Southern Hemisphere (cf. Fig. 26) and the Laurasian countries into the Northern Hemisphere.

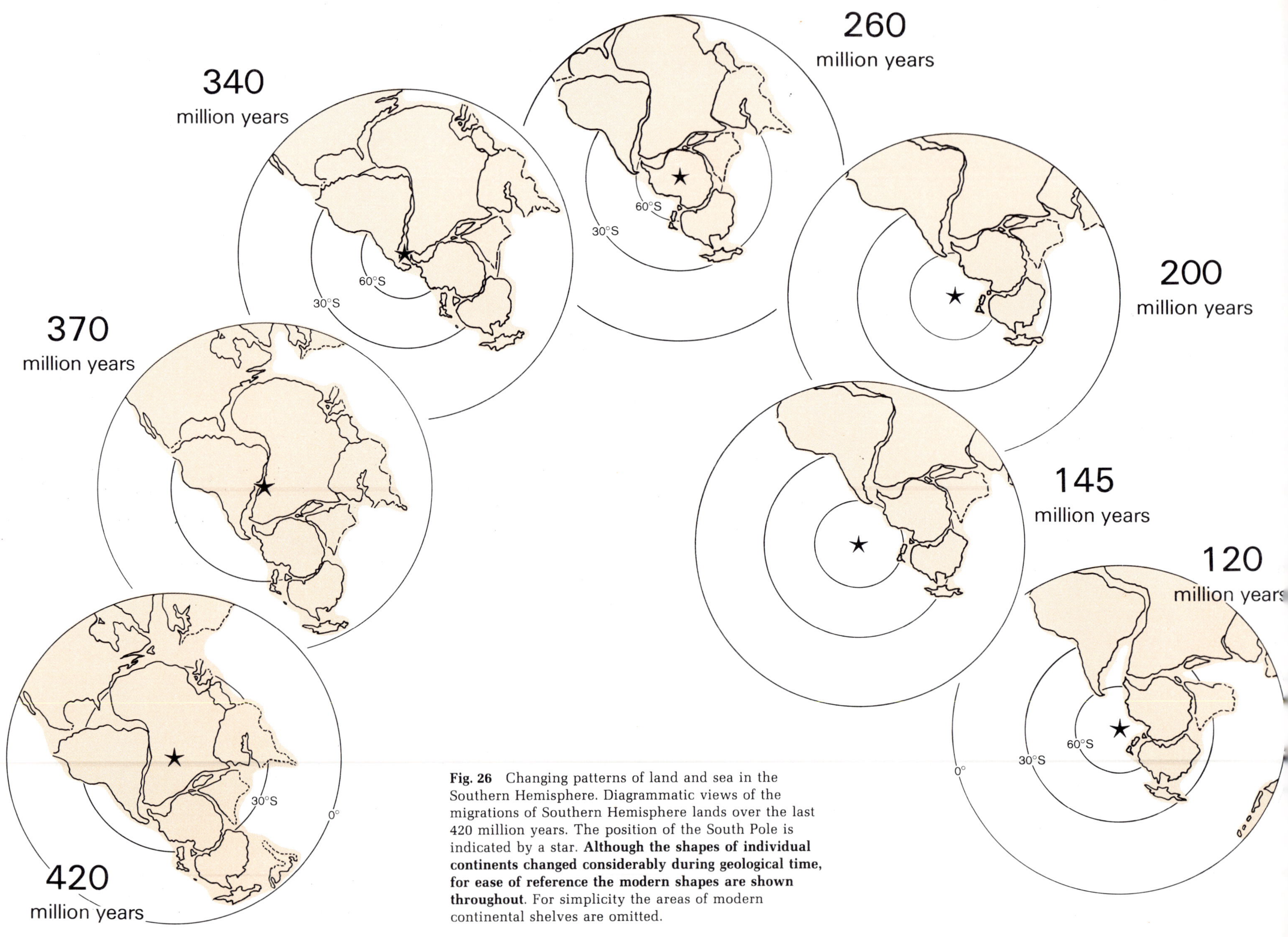

Fig. 26 Changing patterns of land and sea in the Southern Hemisphere. Diagrammatic views of the migrations of Southern Hemisphere lands over the last 420 million years. The position of the South Pole is indicated by a star. **Although the shapes of individual continents changed considerably during geological time, for ease of reference the modern shapes are shown throughout**. For simplicity the areas of modern continental shelves are omitted.

28

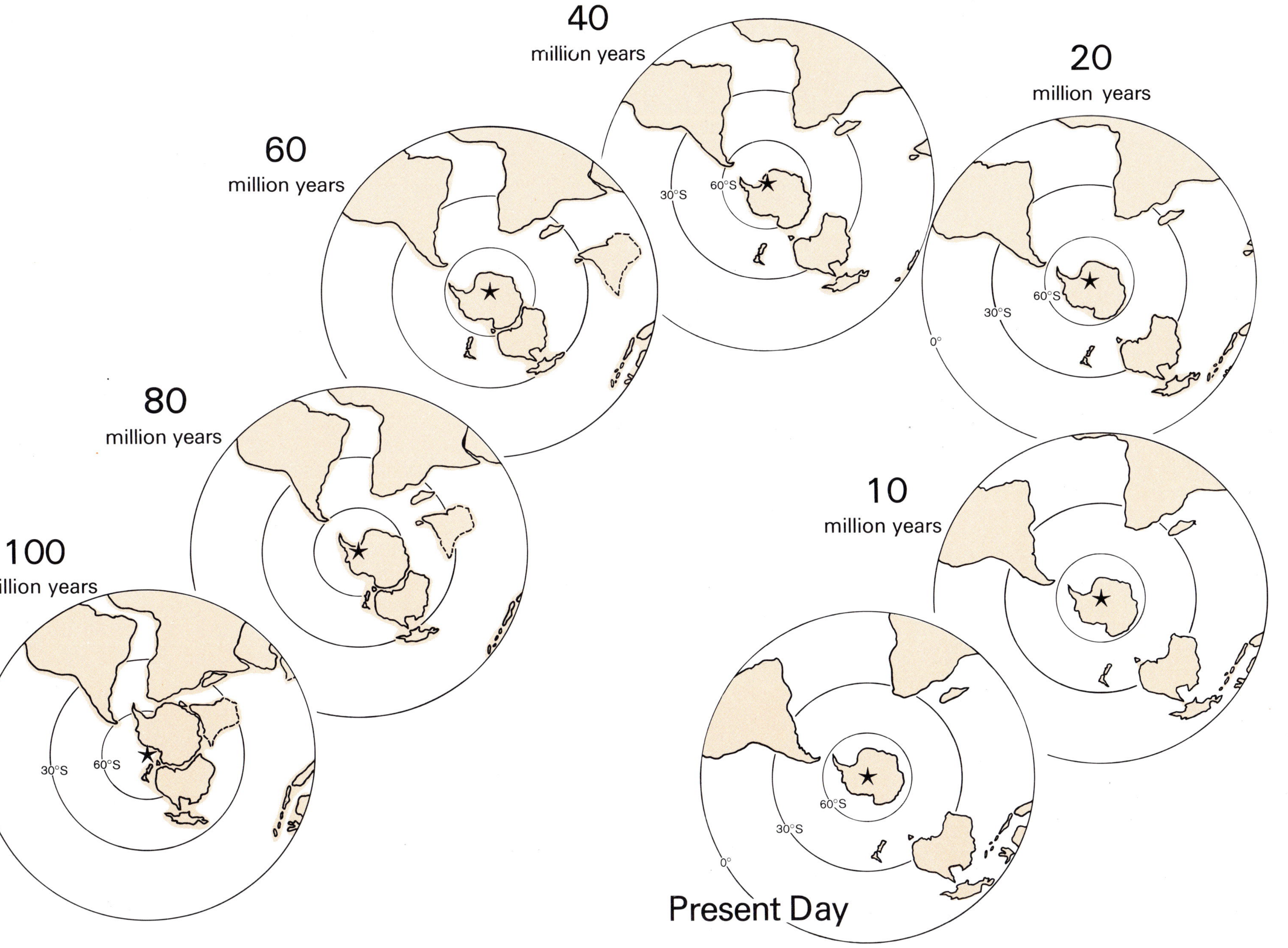
40
million years
60
million years
20
million years
80
million years
100
million years
10
million years
Present Day
30°S
60°S
30°S
60°S
0°
30°S
60°S
0°
30°S
60°S
60°S
30°S
0°

be weak fliers, they can often be blown great distances by storms. (Australian butterflies are regularly blown across the Tasman by westerly storms.) Chance dispersal may also play a major role in the distribution of certain land plants, as exemplified by transport of seeds and living tissue by sea currents or migrating birds. Other plants, however, are extremely sensitive to salt water in all stages of their life cycle (for example the Southern Beech, *Nothofagus*), and would undoubtedly need dry land for migration.

Many coastal marine organisms are, in practical terms, almost as land-bound as most inhabitants of the dry-land. This is because they are restricted in all stages of their life history to the shallow waters of the continental shelves surrounding the land areas. Most of the shelf inhabitants are dependent on light, either to warm the water or to allow the growth of the minute plants and animals on which they feed. Beyond the continental shelves the seas

plunge to depths beyond the reach of light and these deep layers of the oceans generally support different groups of organisms adapted to such conditions. Should a shelf creature or its larvae be swept down into deep oceanic waters, perhaps as a result of a storm, they would probably perish very quickly. Therefore, areas of deep ocean present impassable barriers to shelf creatures. Migration of shelf creatures is possible, however, if individual ocean-bounded lands are linked by continuous areas of shallow shelf seas.

By studying the comings and goings of such environmentally sensitive animals and plants throughout geological time, paleontologists have been able to build up a reasonable picture of which landmass was joined to which, and at what time in geological history. It is also possible to speculate on whether such links were merely areas of shallow seas dotted with islands and archipelagoes, allowing the

migration of shelf marine creatures and perhaps some island-hopping land inhabitants, or whether they also had continuous land enabling a full complement of land animals and plants to cross dry-footed.

The interesting picture that has emerged reflects the fluctuating influence of such links in the past and demonstrates the intimate relationship between faunal and floral migration and the timing of continental movements. At times in the past when the geographic arrangement of the southern continents has been favourable, climates of individual areas reasonably compatible, and requisite migration routes available, New Zealand has been open to a broad range of migrants, both terrestrial and marine. At other times, however, when continental movements have disrupted migration routes and removed or restricted climatic compatability, access to New Zealand was limited.

MAPS OF THE PAST

The following series of maps are called "paleogeographic" maps. They attempt to illustrate the changing geography of New Zealand and neighbouring countries over the past 520 million years, and the influence of such changes on the flora and fauna of the time. Because New Zealand, Australia, Antarctica, and parts of South America, Africa, and India were in the Northern Hemisphere in Cambrian and Ordovician time, Fig. 27 and 29 depict a segment of the Northern Hemisphere. The North Pole is indicated by a star, and the positions of the equator, latitude 30°N and latitude 60°N are indicated by concentric lines.

New Zealand, Australia, and other southern lands moved to positions in the Southern Hemisphere after Ordovician times, so the remaining maps (Fig. 30 onwards) depict a portion of the Southern Hemisphere. On each map the South Pole is indicated by a star and the positions of the equator and latitudes 30°S and 60°S are indicated by appropriately spaced concentric lines.

The shapes of all the land masses varied considerably throughout geological time. New Zealand, for example, only assumed its modern form during the last 10 000 years. At various times broad seaways extended across all of the southern continents as well as those in the Northern Hemisphere and the distribution of land and sea was vastly different from that of today. For ease of recognition, however, the modern outlines of the continents have been used in the following maps and those elsewhere in this book (for example Fig. 3, 6, 10, 25). Examples of how the shape of New Zealand changed throughout geological time are given in Fig. 58 and it must be remembered that equally massive changes have affected most other land masses of the world at various times in their geological history.

Each map incorporates a sample of the life inhabiting the Southwest Pacific at the time, with emphasis on organisms that lived or now live in New Zealand and formed or now form distinctive elements of the flora and fauna. The pattern of diagonal lines denotes a generalised distribution pattern for some of the marine animals living in New Zealand at the time, to give an indication of migration routes available to such animals. Simplified drawings depict representative samples of the New Zealand marine fauna at the time.

Representative land animals and plants are also shown, with arrows indicating possible land routes into New Zealand and New Caledonia from the remainder of Gondwanaland. (Large arrows indicate routes of probable major importance; small arrows minor routes.) The saw-tooth patterns in Fig. 42, 44, 45, 49, and 51 indicate sites of continental rifting around New Zealand, involving disruption of land and sea migration routes.

CAMBRIAN

(570–500 million years ago)   Fig. 27

The period of time before the Cambrian, stretching back to what is thought to be the formation of the earth (about 4600 million years ago), is called the Precambrian Era. During this time complex organic compounds evolved from simple inorganic substances. Simple organisms (bacteria and algae) appeared early in the period, between 3500 and 2500 million years ago. By the end of the Precambrian, soft-bodied marine and freshwater animals such as worms, jellyfish and sponges, and the ancestors of many other animal groups had evolved.

The New Zealand fossil story effectively begins in the Cambrian period. Although patches of older rocks (ranging in age up to 680 million years ago) are known from the West Coast of the South Island (at Charleston and in the Victoria Range near Reefton), little is known about them as they have been squeezed and distorted by earth movements and heated by intrusions of molten rock so that their records of past life (in the form of fossils) have been completely destroyed or made unrecognisable.

Studies of rocks and fossils show that from Cambrian times onwards, and for all of the succeeding geological periods up to about 80 million years ago, New Zealand was positioned at the eastern edge of the super-continent Gondwanaland, wedged between Victoria (Australia) and Victoria Land (Antarctica), and that this basic relationship did not alter to any marked degree. The geography of the world at that time was vastly different from that of today (see Fig. 24–26). All of the lands now lying in the Southern Hemisphere (i.e., South America,

Africa, Madagascar, Australia, Papua New Guinea, New Zealand), together with India, were grouped together into a single landmass (Gondwanaland). On the other hand, most of the lands now lying in the Northern Hemisphere, were fragmented and widely distributed across the globe. For example, Asiatic U.S.S.R., Southeast Asia, China, Mongolia and Manchuria, later to become components of Eurasia, were separate landmasses and, at least in the Cambrian and succeeding Ordovician and Silurian periods, much closer to Australasia than today.

The available evidence suggests that throughout Cambrian times virtually all of Gondwanaland was situated in middle and low latitudes of the Northern Hemisphere, whereas large areas of Eurasia were situated in the Southern Hemisphere— almost the exact opposite of the modern situation! New Zealand in the Cambrian was sited at a latitude of 40–50°N, and, along with Australia, was close to a landmass that included areas now comprising modern Southeast Asia, China, Japan, and Korea (see Fig. 24–26).

Throughout most of Cambrian time shallow seas covered many areas of Australia and Antarctica. As far as we know New Zealand was also largely under the sea. At times the seas covering New Zealand were dotted with volcanic islands, and submarine lava flows, silts, and limestones were laid down on the sea bed around these islands.

The marine organisms that populated New Zealand in the Cambrian had very close

links with Australia, but at the same time links also existed with Southeast Asia, China, Mongolia, Manchuria, Asiatic Siberia, Tien Shan, and the Central Asian regions of U.S.S.R. (i.e., Kazakhstan, Turkmenistan, Uzbekistan, Tadshikistan, and Kirgizstan).

Australia and Antarctica were in tropical latitudes during most of Cambrian time. It is therefore not altogether surprising that warm-water reef-building organisms (Archaeocyatha), similar to the modern corals, are found in the Cambrian deposits of both countries.

Judging from its mid-latitude position, New Zealand in the Cambrian probably had a warm-temperate climate, very much like that of today. A warm-temperate climate is also indicated both by the affinities of the Cambrian fossils found in New Zealand and by the features of the rocks laid down at the time (notably the absence of any indications of extremes of climate, that is, hot or cold). The absence of warm-water reef-building Archaeocyatha from New Zealand, for example, indicates that New Zealand was outside the tropical regions favoured by such organisms.

Glaciations had occurred many times in the Precambrian, extending up to 670 million years ago. However, the Cambrian appears to have been a period when the world was free of ice, probably because no significant landmass was within 30° of either North or South Pole. It is likely that the tropical, subtropical, and temperate zones were wider than those of today and world climate was correspondingly more equable.

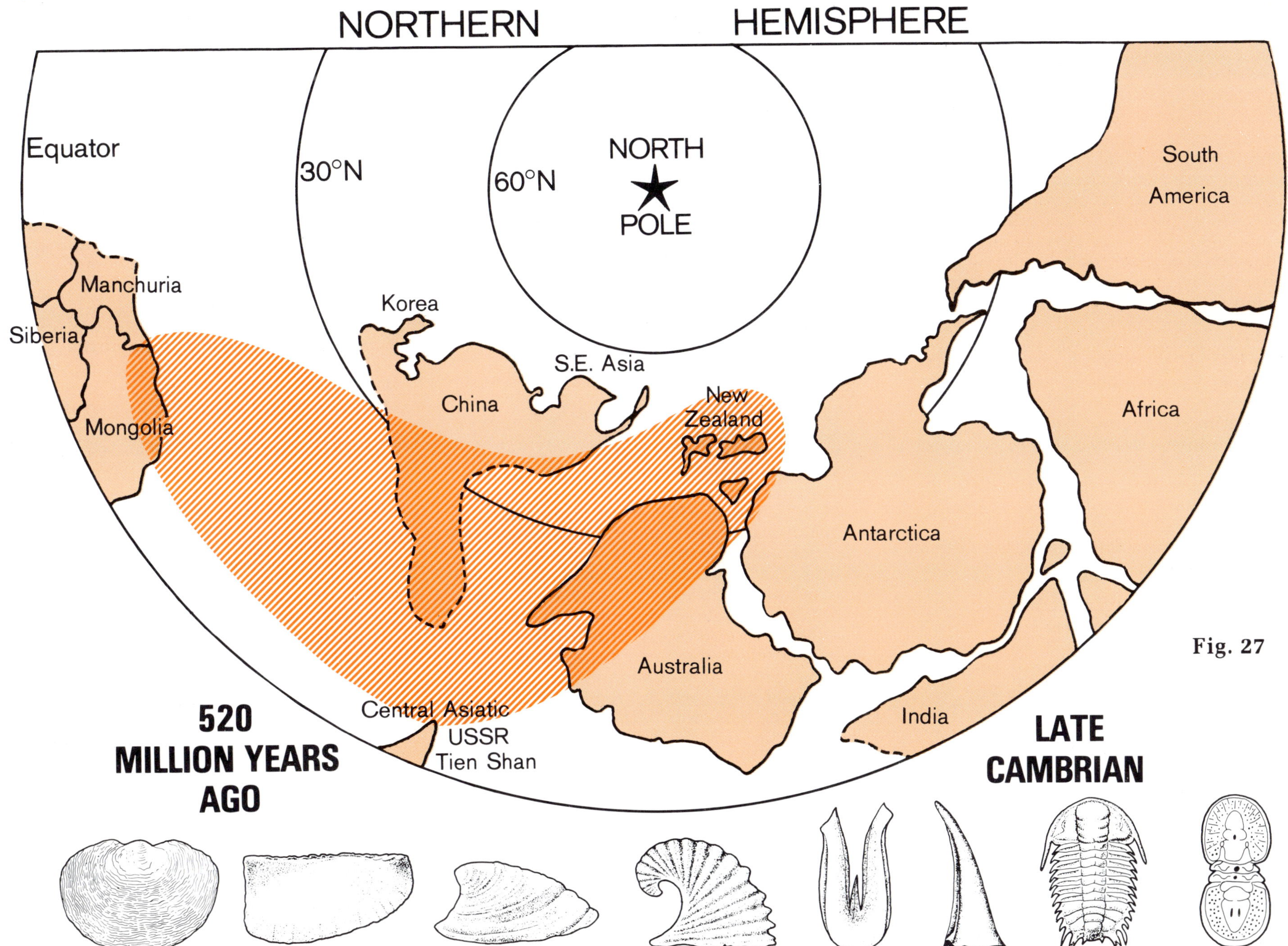

NORTHERN HEMISPHERE
Equator
30°N
60°N
NORTH POLE
Siberia
Manchuria
Mongolia
Korea
China
S.E. Asia
New Zealand
Central Asiatic USSR Tien Shan
Australia
Antarctica
India
Africa
South America
520 MILLION YEARS AGO
LATE CAMBRIAN
Fig. 27

CAMBRIAN LIFE

At the beginning of the Cambrian, shells and hard parts were developed by a number of animal groups, including brachiopods, sponges, and arthropods. Brachiopods, sometimes called "lamp shells" as they resemble Roman oil lamps, are a type of shell fish, with two shells or "valves", but, unlike the clams (or bivalves), the valves are carried above and below, rather than left and right. The brachiopods continued throughout geological time, becoming a most important and diverse fossil group, although they are rare in modern seas.

The Arthropods (jointed-legged creatures), which include the modern crayfish, insects, millipedes, and centipedes, were represented in Cambrian times by the aquatic trilobites, named from the fact that their bodies were divided into three lobes. These distinctive and often abundant creatures, many of which superficially resembled modern wood-lice, dominated seas for the next 100 million years, becoming extinct at the close of the Permian period, 235 million years ago.

Ancestors of spiny-skinned creatures (Echinoderms) were represented in the Cambrian. They were later to give rise to sea urchins and sea eggs, starfish, sea lilies, and sea cucumbers. Also appearing in the Cambrian period were the first representatives of Mollusca— a group that later expanded to include animals such as chitons (Amphineura), tusk shells (Scaphopoda), clams, mussels, scallops (Bivalvia), whelks, snails, slugs, limpets (Gastropoda), squids, cuttlefish, and ammonites (Cephalopoda). Conodonts also appeared in the Cambrian. They are microscopic phosphatic fossils resembling delicate fang-like teeth. The Conodont animal was soft-bodied and rather like the modern jawless hagfish (lamprey). It probably lived on the sea floor, and fed by sucking in or scooping up surface mud and organic detritus. It is not known whether the conodonts functioned as true teeth, or provided support for gills.

WHY SHELLS??

Precambrian animals, preserved as fossils in rocks 700–680 million years old, include sea pens (pennatulaceans) and jellyfish. Rocks 620–580 million years old contain sea pens, jellyfish, sponges, and various marine worms. Other fossils, harder to identify, may be ancestors of molluscs, arthropods, and echinoderms. All these fossils give the impression of being almost entirely soft-bodied, although some may have been stiffened by spines, ribs, or layers of resistant materials.

Hard parts (skeletons, shelly coverings, etc.) first appear in abundance in the earliest part of the Cambrian, 570 million years ago. Then, over the next 30 million years most of the organisms preserved as fossils develop hard parts, primarily based on calcium/magnesium carbonate or phosphate. Ten or more invertebrate groups (phyla) including brachiopods, molluscs, sponges, arthropods, echinoderms, and conodonts, together with algae and protistans (foraminifera) developed hard parts in the earliest Cambrian, and corals followed in the middle Cambrian.

Some paleontologists view development of hard parts as simply a step in the evolution of life, needing no special explanation. Others, impressed by the sudden development of such a diverse range of organisms, feel that some chemical and environmental factors acted as trigger mechanisms. Chemical factors include oxygen levels in the atmosphere. The components used to build hard parts in marine creatures are extracted from the sea. Such extraction processes require specific chemical and physical conditions in both the sea and atmosphere. The level of oxygen is one such condition. It has been proposed that the original atmosphere of the earth lacked oxygen and that the build-up to the present level of 21% is entirely due to the activities of plants, notably marine algae. By the beginning of the Cambrian oxygen levels had reached such a point to allow the synthesis of collagen, an essential component of both hard parts and large muscles.

Environmental factors include better levels of nutrition (resulting from improved oxygen content), development of active predators, and the opening up of new ecological niches as continents moved.

CAMBRIAN IN NEW ZEALAND

New Zealand's Cambrian fossils are found chiefly in limestones that were laid down on the sea bed, around volcanic islands and archipelagoes offshore from the coastline of eastern Australia–Tasmania–Antarctica. Only a very small slice of Cambrian time is preserved in the New Zealand fossil record: about 25% of total Cambrian time, representing parts of the middle and late Cambrian. Nonetheless, the fossils present in New Zealand are a good cross-section of the faunas living at the time. They comprise sponges, brachiopods, molluscs, trilobites, and conodonts (Fig. 28). Cambrian fossils are known from Cobb Valley (northwest Nelson) and Springs Junction (west Nelson).

Representative New Zealand Cambrian fossils are illustrated on Fig. 27 (left to right): the brachiopod *Micromitra*; the bivalve *Tuarangia*; molluscs *Mellopegma* and *Latouchella*; conodonts *Westergaardodina* and *Hertzina*; and trilobites *Dorypyge* and *Ptychagnostus*.

Fig. 28 A glimpse of New Zealand sea floor life in the middle Cambrian, 540–525 million years ago. At this time New Zealand was part of the sea floor surrounding strings of volcanic islands (diagram, top right). The islands lay some distance offshore from continental land, now forming parts of Australia and Antarctica. The animals living on the sea floor were mainly small, with the largest being the trilobite *Dorypyge* (centre) up to 80 mm in length. Other animals depicted are conical shelled hyolithids, related to molluscs, *Ptychagnostus* a small trilobite up to 10 mm in length, and small lamp shells (brachiopods). The sea floor is littered with the hard skins of trilobites that have been shed during stages of growth (as modern cicadas shed their skins). Foraging in the sea floor muds are conodont animals, soft-bodied organisms thought to be similar to the modern lamprey and feeding by sucking in or scooping up organic material contained in the mud. A group of sponges are attached to the rocks (left), and a jellyfish hovers above the scene (top left).

ORDOVICIAN (500–430 million years ago) Fig. 29

During the Ordovician period all of the world's landmasses were affected by gradual rotating movements. Towards the end of Ordovician times, New Zealand, Australia, and other Gondwana countries had swung into the Southern Hemisphere and most of the Laurasian countries into the Northern Hemisphere. Such movements also swung the land fragments of Central Asiatic U.S.S.R. and Tien Shan, China, Manchuria, Mongolia, Asiatic Siberia, and Southeast Asia northwards away from Australasia, thus loosening and eventually breaking the marine links formerly existing between these countries and Australasia.

In the middle Ordovician (illustrated), gradual rotating movements of the world's landmasses had swung New Zealand southwards to a position of 15–20°N latitude (compared with 40–50°N in the Cambrian).

Throughout the Ordovician, New Zealand marine organisms had very strong links with eastern Australia but, as in the Cambrian, links also extended further afield to Manchuria, Mongolia, Asiatic Siberia, China, Southeast Asia, Central Asiatic U.S.S.R., and Tien Shan. Marine links also extended to South America and around the margins of the infant Pacific Ocean to western North America (parts of which were further south than today).

Judging from its geographical position, New Zealand had a tropical/subtropical climate for at least some of Ordovician time. This supposition is supported by the general affinities of marine life of the time and also by the presence of ancient reef corals (presumably tropical organisms like their modern counterparts).

The early and middle Ordovician, like the Cambrian, appear to have been largely free of ice, as no landmass was within 30° of either North or South Pole. Later in the Ordovician, as continued rotation of the world's landmasses swung various lands closer to the poles, evidence of ice action became apparent. By the end of the middle Ordovician local centres of glaciation developed in Scotland, Ireland, Normandy, Spain, and Portugal. In the late Ordovician extensive continental ice sheet glaciation occurred in north and west Africa.

ORDOVICIAN LIFE

Animal life flourished in Ordovician seas. Especially prolific were trilobites, brachiopods, and the three mollusc groups of bivalves, gastropods, and cephalopods. The first coral reefs and colonies of moss-like bryozoa appeared during the Ordovician. The first traces of fish are also known from the Ordovician, although only fossilised scales are known. Graptolites also became abundant in some marine environments. These were colonial floating animals some of which looked like branching spider webs. Graptolites are thought to be primitive Protochordates, that eventually gave rise to the vertebrates.

ORDOVICIAN IN NEW ZEALAND

In contrast to the Cambrian, the New Zealand Ordovician record is almost fully representative of the entire span of Ordovician time. The most important and most complete Ordovician localities are situated in Northwest Nelson (notably Aorangi Mine, Cobb Valley, Mount Patriarch, Takaka Valley, Wangapeka Valley). Other localities are in the Baton River (west Nelson) and at Cape Providence and Preservation Inlet (Fiordland).

Like the record of the rocks, the Ordovician fossils preserved in New Zealand are also fully representative of marine life of the time and include corals, brachiopods, trilobites, crustacea, conodonts, and very abundant graptolites. Most of these fossils had very close affinities with eastern Australia. Less strong, but still quite marked affinities were also apparent with Asiatic U.S.S.R., China, Siberia, Manchuria, Mongolia and, continuing around the northwestern edge of the Pacific, with western North America. Two of the New Zealand Ordovician trilobites also show affinities extending along the southeastern edge of the Pacific, to South America.

Drawings of some of the more abundant New Zealand Ordovician fossils are placed along the lower margin of Fig. 29 (from left to right): the graptolites *Tetragraptus*, *Isograptus*, and *Oncograptus*; the trilobite *Incaia*; and the conodonts *Pygodus* and *Eoplacognathus*.

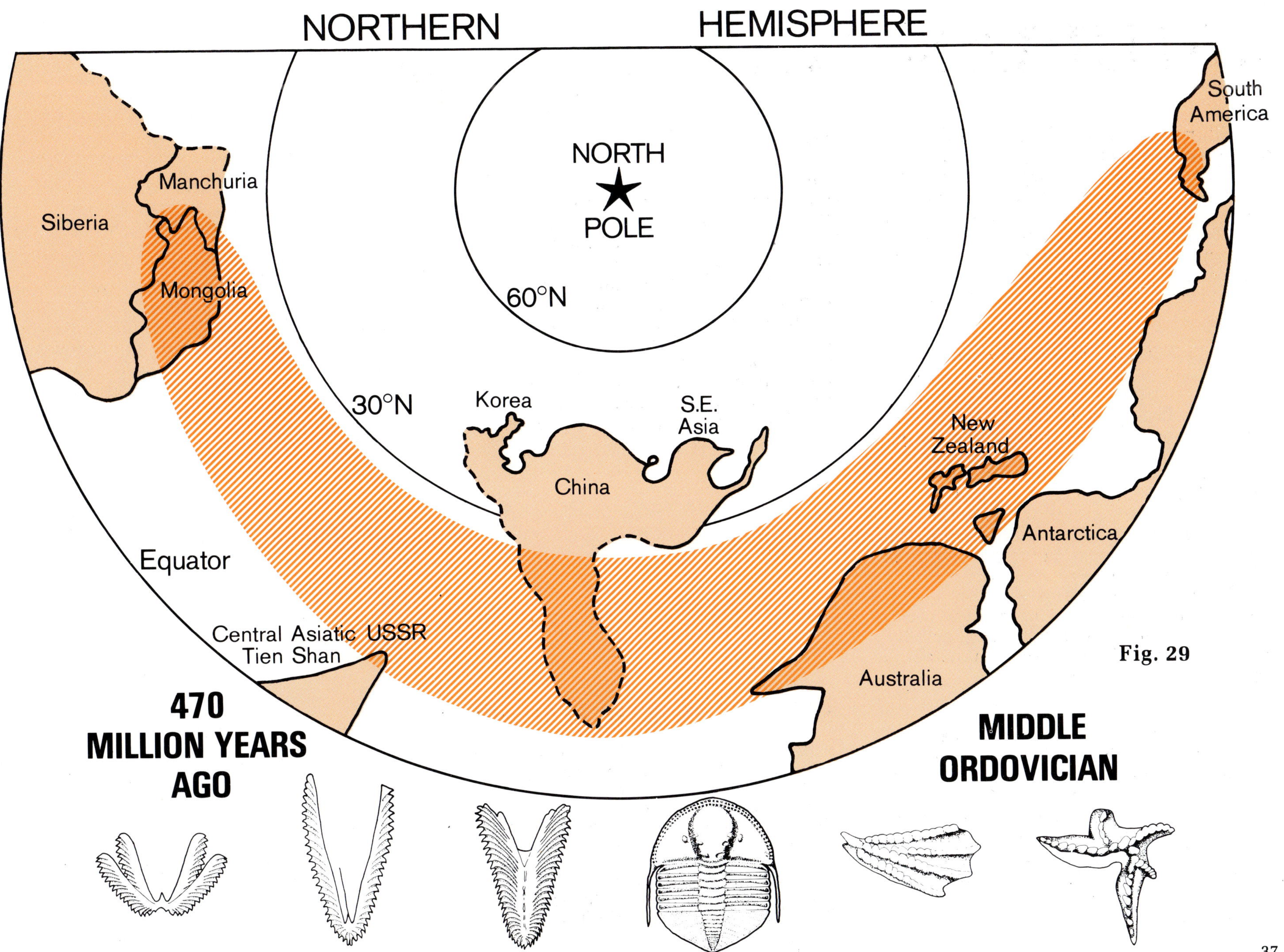

NORTHERN HEMISPHERE
NORTH POLE
60°N
30°N
Equator
Siberia
Manchuria
Mongolia
Korea
China
S.E. Asia
Central Asiatic USSR Tien Shan
New Zealand
Australia
Antarctica
South America
470 MILLION YEARS AGO
MIDDLE ORDOVICIAN
Fig. 29

SILURIAN

(430–400 million years ago) Fig. 30

The Silurian period saw a continuation of the rotational movements affecting the world's landmasses, that had progressively moved Australia and New Zealand from positions in middle and low latitudes of the Northern Hemisphere in Cambrian and early/middle Ordovician times (Fig. 27, 29) into low latitudes of the Southern Hemisphere by the close of Ordovician time.

By middle Silurian times (illustrated) New Zealand had moved to a geographic position of 25–30°S latitude and Gondwanaland as a whole was beginning to straddle the South Pole. In response to the gradual rotation of the Gondwana landmass across the South Pole, the focus of glaciation shifted from its Ordovician position in northern and northwestern Africa, southwards to the Sierra Leone coast and southern Africa. This can be seen by comparing the pole positions in Fig. 24–26. Gradual rotation of Gondwanaland southwards was also responsible for the commencement in early and middle Silurian times of glacial activity in various areas of South America, notably Brazil.

The rotation of New Zealand southwards into the Southern Hemisphere broke the faunal links with Mongolia, Manchuria, Siberia, and Central Asiatic U.S.S.R. that were such distinctive features of Cambrian and Ordovician times. Strong links continued with eastern Australia, and weaker links continued with Southeast Asia and China, which had retained their positions relative to New Zealand.

SILURIAN LIFE

The Silurian was a time of major expansion for many of the marine invertebrate groups that had first appeared in the Cambrian and Ordovician. Trilobites, sea lilies (crinoids), corals, brachiopods, graptolites, and conodonts all became highly diverse and expanded to occupy a great range of marine environments. Although corals had first appeared in significant numbers in the Ordovician, the first coral reefs appeared in the Silurian. They are often preserved in the rocks as massive beds of limestone. The first fish with jaws (placoderms) appeared in the seas of latest Silurian times. Small fish and large arthropod water-scorpions (eurypterids) abounded in fresh-water lakes, rivers, and streams.

Judging from the fossil record preserved in the rocks, the latter parts of the Silurian saw the first colonisation of the land by primitive plants. However, although some water-scorpions may have ventured onto dry mud flats edging lakes and rivers, the real conquest of the land by animals had to wait until the Devonian.

SILURIAN IN NEW ZEALAND

In New Zealand we see only a glimpse of Silurian life. This time was an episode in New Zealand's history characterised by massive earth movements and volcanic disturbances (the Greenland Event), leaving little chance for preservation of the fossil record.

Silurian fossils are known from two localities in northwest Nelson: Pikikiruna Range and Wangapeka Valley. The fossils include six types of brachiopods, representing a slice of Middle Silurian time, spanning about 14 million years. Their affinities are closest with eastern Australia, but links further afield extend towards China and Southeast Asia.

Two of the Silurian brachiopods found in New Zealand are illustrated in the top right corner of the diagram: *Conchidium* (upper) and *Isorthis* (lower).

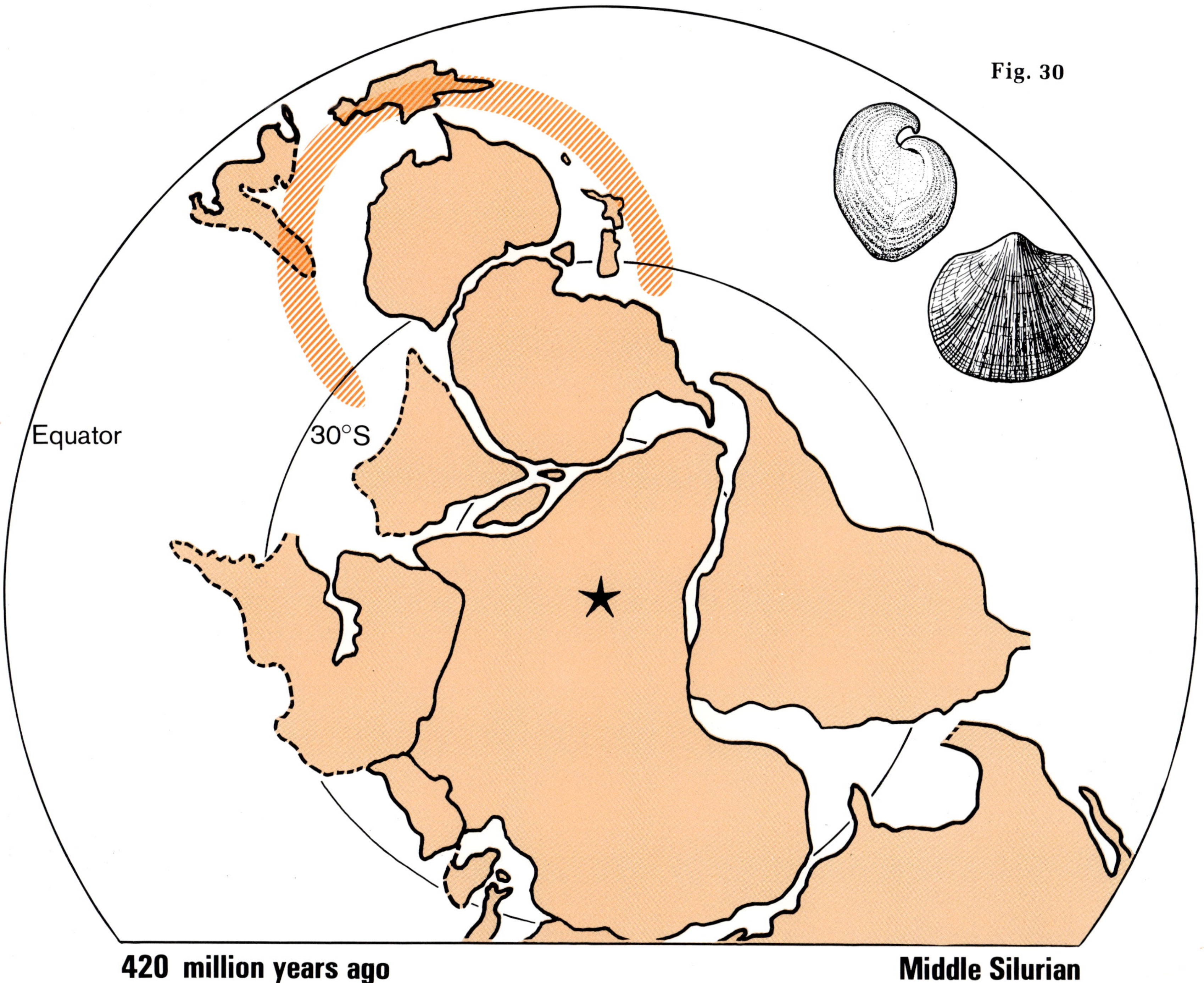

Fig. 30
Equator
30°S
420 million years ago
Middle Silurian

DEVONIAN (400–350 million years ago)   Fig. 31

The progressive southward drift of the Gondwana countries seen in the preceding Cambrian–Silurian times (Fig. 27, 29, 30) also continued throughout the Devonian. As a consequence, New Zealand was slightly further south in the Devonian than in Silurian times, and had a geographic position of 25–35°S latitude.

The changing position of Gondwanaland with respect to the South Pole was also reflected in the records of glacial activity. In the late Devonian glacial activity was evident in the Niger Basin of west Africa and the Amazon and Parnaiba Basins of Brazil, with some activity in southern Chile.

As more and more of Gondwanaland came to lie in southern latitudes and was affected by the cooling effects of polar ice, distinctively "southern" links appeared in marine faunas of the time, presumably related to cool-temperate climatic conditions then prevailing around the South Pole. New Zealand participated in this trend. Although very strong links continued with Australia, the Asiatic affinities evident in Cambrian–Silurian times disappeared completely in the Devonian. As the Asiatic continental blocks moved northwards into the Northern Hemisphere, New Zealand moved southwards into the southern temperate zone. New Zealand marine faunas assumed links with South Africa, Bolivia, and Antarctica, indicating establishment of shallow-water marine migration routes around the eastern margin of Gondwanaland.

DEVONIAN LIFE

Devonian marine faunas differed quite significantly from those of the Silurian. Trilobites, so abundant in pre-Devonian ages, became very much less common in the Devonian. Graptolites also went into a decline. However, brachiopods, cephalopods, and various groups of clams continued to evolve and expand. Reef corals, moss-like bryozoa, and sponge-like stromatoporoids flourished in tropical seas.

The Devonian, often called the "Age of Fishes", saw a great expansion of fish of all types in both the sea and freshwater. Sharks and spiny fish appeared and rapidly colonised many regions of the world. Large fish, some covered with massive armoured plates or thick scales populated marine and fresh waters. Some of the scaly fish, with thick lobe-like fins (ancestors of the modern coelacanth) began to venture onto mud flats and eventually gave rise to the first amphibians.

The Devonian is also called the "Age of Ferns", as many developments of land plants took place at this time. The simple plants that had first colonised the land in the late Silurian were followed in the Devonian by club mosses, horsetails, and a great variety of ferns. Many reached a considerable size and formed large forests.

DEVONIAN IN NEW ZEALAND

Although there is no record of land plants in the New Zealand Devonian, nor of freshwater life, there is a reasonably comprehensive sample of marine life. New Zealand Devonian faunas include bivalves, brachiopods, trilobites, corals, stomatoporoids, byozoa, crinoids, primitive echinoderms, conodonts, and tentaculites (small conical tube-like fossils, possibly related to bryozoa). Several of these fossils show southern affinities, linking New Zealand and Australia with Antarctica, South America, and South Africa. Three fossils with such affinities are illustrated in Fig. 31 (clockwise from the top): the brachiopods *Maoristrophia* and *Acrospirifer* ; the trilobite *Burmeisteria (Digonus)*. Some Devonian fossils are of cosmopolitan groups that achieved a very wide geographic distribution. Others were endemic to New Zealand, and show the beginning of some degree of geographic isolation (a trend greatly reinforced in later geological times). Devonian fossils are known from three areas: near Reefton, Baton River (110 km north of Reefton), and near Lake Haupiri, north Westland.

Although the New Zealand Devonian faunas are numerically rich, they represent only about one half of the total Devonian time; that is, parts of early and middle Devonian, spanning about 27 million years. A large slice of later Devonian time is missing. This time coincided with the beginning of major earth movements in New Zealand— the Tuhua mountain building episode. These movements created areas of broken rugged landscape and many active volcanoes dotted the land and sea. For this reason the marine record is virtually non-existent for both the late Devonian and the succeeding Carboniferous.

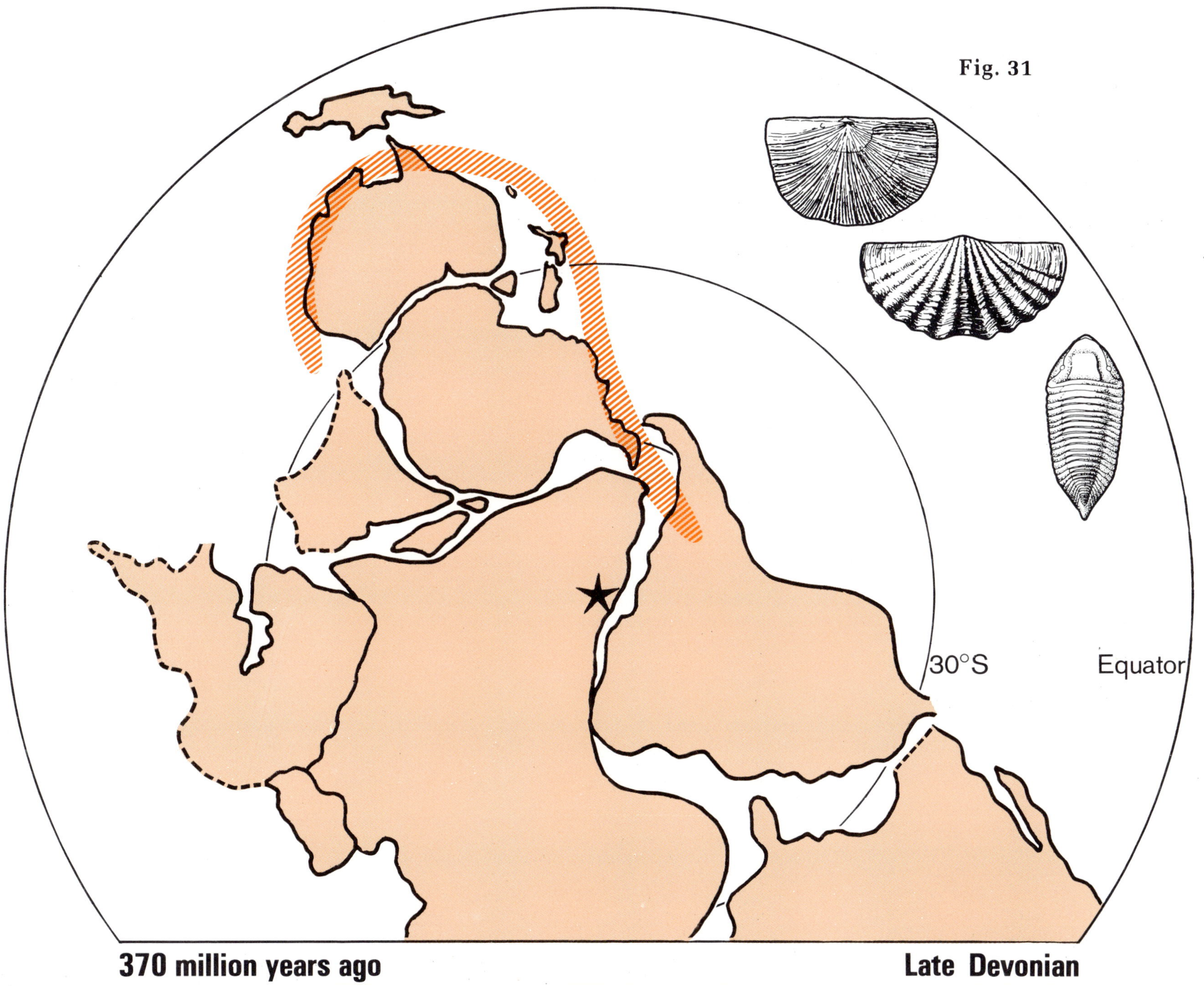

Fig. 31
30°S
Equator
370 million years ago
Late Devonian

CARBONIFEROUS (350–300 million years) Fig. 32

Carboniferous times saw the culmination of the steady drift of Gondwanaland towards the South Pole. Gondwanaland virtually straddled the South Pole. Coincident with the southwards drift, mountain building movements affected much of eastern Gondwanaland in late Devonian and early/middle Carboniferous time. In the south polar region, these earth movements created substantial areas of mountainous land, which excluded the moderating influence of the sea. The presence of this high country close to the South Pole generated stormy weather conditions and formed a focus for the precipitation of snow and ice. The valleys and depressions of the high country also provided concentrating grounds for the snow and ice, where it became compacted to form hard glacier ice. Once formed, the glacier ice flowed outwards, under the influence of gravity, covering the surrounding lands and pushing out across the sea as floating ice sheets.

The effects of the southern drift of Gondwanaland, combined with episodes of mountain building, generated a series of southern glacial periods affecting large parts of Gondwanaland. These began in the late Carboniferous and continued into the succeeding Permian. During these times large continental ice sheets scraped and scoured many areas of southern South America, southern Africa, and India in the Carboniferous, and Australia and Antarctica in the Permian. In the early Carboniferous, as illustrated, New Zealand was at a latitude of about 50°S and was probably subject to cool- or cold-temperate climatic influences.

CARBONIFEROUS LIFE

Away from the cold climates, the Carboniferous was a time when many plants and animals consolidated their places in the environments of the time.

On land, plant life developed a rich assemblage of giant tree-ferns, horsetails, clubmosses, lycopods, seed ferns, and cone-bearing plants, that grew luxuriantly in the tropical and subtropical belts of the world. The forests produced quantities of woody plant debris which accumulated in swamps as thick layers of peat. Burial of the peat beneath later sediments slowly transformed the peat into coal. Coal deposits of North America and Europe, Asia, and Australia were formed in this way. No coals accumulated in New Zealand at the time because little, if any, land was present. The main deposits of coal in New Zealand are very much younger: late Cretaceous, Paleocene, and Eocene.

In the sea, many of the old lines of invertebrates (including many, but not all graptolites and trilobites) became extinct, or greatly reduced, at the end of the Devonian.

These extinctions were followed in the early Carboniferous by a new phase of evolution and expansion of marine animals. The wide shallow shelf seas around the continents provided abundant and secure habitats for shelled creatures. Corals, brachiopods, bryozoa, cephalopods, and bivalves abounded, and in some areas dense masses of sponges and sea lilies (crinoids) formed groves on the sea floor. Their skeletons later formed limestone deposits. Seaweeds were also widespread, and some produced thick calcareous crusts on the sea floor. Sharks dominated the sea. Crustaceans, such as lobsters and crabs, took over the place in the environment vacated by the Trilobites. They became the main scavengers of the sea bed. They also evolved means of protection from enemies by developing armour plating and a formidable array of spines and claws, and by constructing complicated living galleries within the sea floor sediment.

Amphibians became a large and important group and, towards the end of the Carboniferous gave rise to the reptiles— the first true land vertebrates. Insects evolved rapidly during the Carboniferous and large dragonflies lived in many forests.

CARBONIFEROUS IN NEW ZEALAND

Throughout the Carboniferous the New Zealand area was largely occupied by the landmass formed by the Tuhua earth movements in late Devonian and early Carboniferous times. Consequently, there is very little of a Carboniferous marine record preserved in New Zealand and only one locality is known: Kakahu, South Canterbury. The only fossils from Kakahu are microscopic conodonts and fragments of fish scales. Two specimens of conodonts are illustrated (upper right margin) in Fig. 32: *Cavusgnathus* (above) and *Idiognathodus* (below). Judging from these fossils, and the more numerous Carboniferous faunas of Australia, marine links such as those depicted on the map existed around the Indo-Pacific margin of Gondwanaland for most of Carboniferous time.

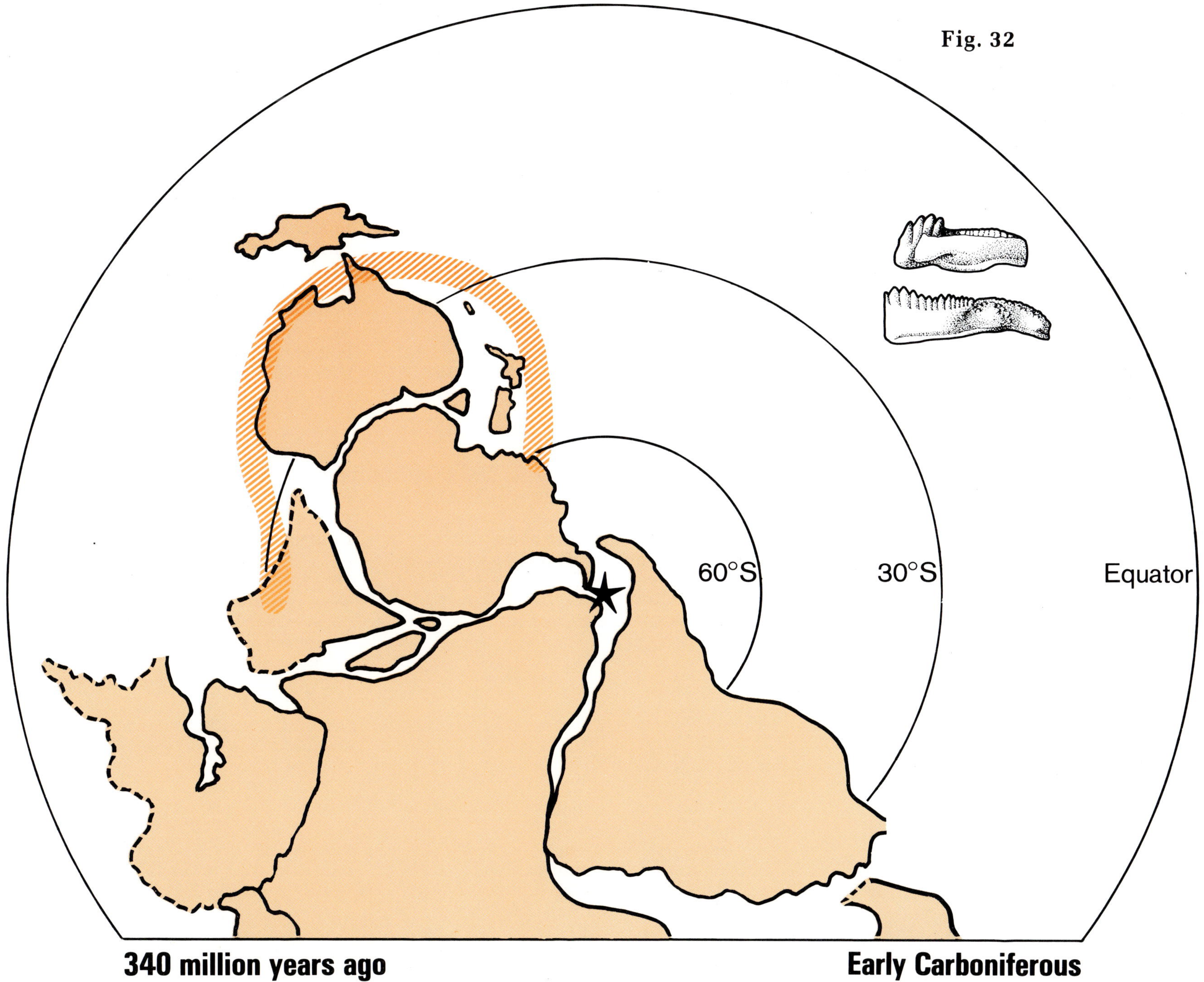

Fig. 32
60°S
30°S
Equator
340 million years ago
Early Carboniferous

PERMIAN

By Permian times, the slow drifting of Gondwanaland across the South Pole had progressively moved the centre of glaciation from southern Africa to Antarctica (compare Fig. 31–33). As a consequence, the effects of ice action are more evident in Permian times in Australia and Antarctica, whereas in the Carboniferous the major effects of glaciation were felt in southern Africa and South America.

The southerly drift of Gondwanaland had moved New Zealand closer to the South Pole so that in middle Permian times its position was 60–70°S latitude. Although New Zealand was under the sea for much of Permian time, and so probably escaped the effects of continental glaciation, it nonetheless experienced cold climates and at these times its seas were probably dotted with icebergs. During these glacial ages the Australian and New Zealand seas were populated by cold water marine organisms.

Although ice covered much of the land in the Gondwana continents during the glacial ages, its distribution was patchy and numerous ice-free areas existed around the margins of individual ice sheets. These areas were populated by seed ferns called *Glossopteris*. *Glossopteris* is not known from Laurasia and occurs only in rocks of Permian age in the Gondwana countries, where it is often found in close association with glacial deposits. The general pattern of its distribution is shown by the symbols on Fig. 33. Until recently *Glossopteris* was unknown in New Zealand, leading to the possibility that at least in the Permian New Zealand was not part of Gondwanaland.

However, as work progressed on the New Zealand Permian rocks the reason for such an absence became apparent. The New Zealand Permian rocks were predominantly of marine origin, therefore the chances of finding the remains of land plants were rather slender. However in 1970 a single leaf of *Glossopteris* was found in Southland. This leaf probably drifted off-shore from the main landmass of Gondwanaland and, when it became water-logged, settled on the sea floor, to be incorporated in sands and muds.

The discovery of this fossil provided firm links between New Zealand and the remainder of Gondwanaland and gave some indication of the closeness of New Zealand to land in the Permian (Fig. 34). Although actual land routes may also have existed, the land created in New Zealand at the time was largely volcanic and, therefore, ephemeral. The direction of likely land routes are indicated by arrows on the map.

PERMIAN LIFE

One of the major consequences of the build-up of ice in Gondwanaland in late Carboniferous and Permian times was development of sharply differentiated climatic belts, similar to those of today, but somewhat broader. Each belt, like its modern counterpart, had its distinctive flora and fauna and, although minor subdivisions are possible, these roughly corresponded to tropical, subtropical, warm temperate, and cold temperate. The sharpest differentiation of these belts occurred in Permian times, followed by reversion to greater uniformity in the succeeding Triassic period.

In the seas of all the Permian climatic belts, brachiopods, bivalves, and primitive bony fish continued to thrive. Cephalopods, while abundant in tropical and subtropical seas, occurred sparsely in cooler seas. Compared with Carboniferous seas, Permian seas had fewer corals, sea-urchins, and sharks.

Fusulinids, a group of single-celled microscopic animals (Foraminifera) that had appeared in Carboniferous times, became very abundant in the Permian, particularly in tropical and subtropical seas.

On land, reptiles steadily gained supremacy over amphibians as Permian time progressed. Squat Permian reptiles, heavily armoured and some attaining a height of 1.5 m and length of 2.5 m lumbered over many lands, especially those situated in the tropical and subtropical zones of the time (including southern European Russia, southern Europe, Texas, Oklahoma, and New Mexico).

Reptiles did not penetrate very far into Gondwanaland. The reason for this was simply that many areas of Gondwanaland were too cold for cold-blooded creatures. Throughout the Permian, cold climates undoubtedly excluded land reptiles from Antarctica and Australia (the focal points for Permian glaciation). Towards the end of the Permian, however, land reptiles had moved

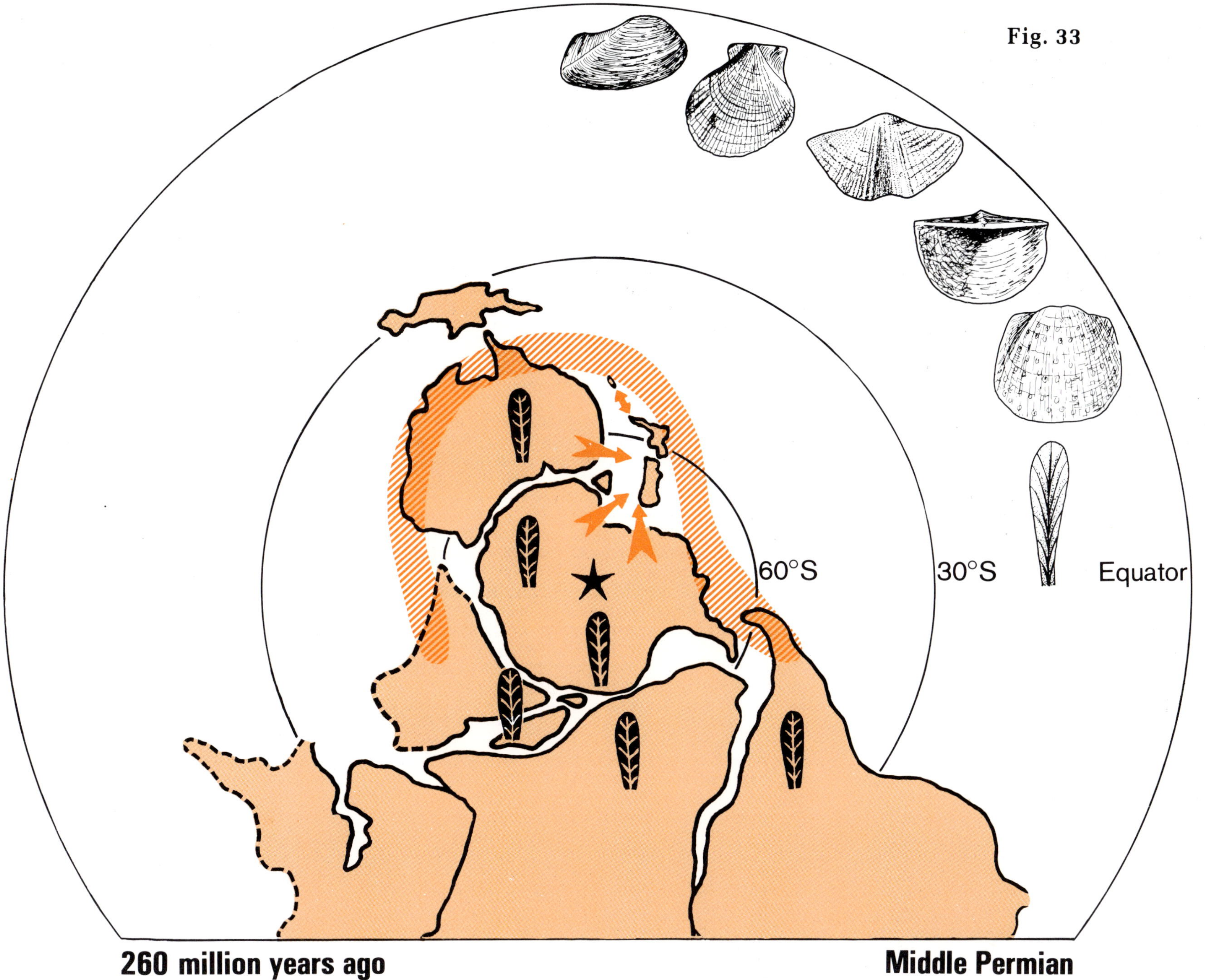

Fig. 33
60°S
30°S
Equator
260 million years ago
Middle Permian

into areas of Gondwanaland that had been glaciated in the Carboniferous, but were largely free of ice in Permian times. These areas included southern and eastern Africa, Brazil, and northern India.

The Permian was also a time of expansion for other land inhabitants. Butterflies, moths, and flies appeared and rapidly became numerous. The floras of southern lands were dominated by the seed fern *Glossopteris* (Fig. 34). Elsewhere giant tree ferns, horsetails, club mosses, and lycopods grew in vast forests. Cone-bearing plants were also present, but in a minority.

PERMIAN IN NEW ZEALAND

The landmass created by the Tuhua earth movements during late Devonian and early Carboniferous times was largely eroded away by early Permian time. New Zealand then entered a period of almost 165 million years, from Permian to Jurassic times, when sea occupied most of its area. During this time, however, New Zealand was never very far away from a shoreline: either adjacent to continental areas of Australia, Tasmania, and Antarctica, or fringing ephemeral small islands or archipelagoes thrown up by local earth movements. The seas covering New Zealand received sediments from the erosion of Australia, Tasmania, and Antarctica, as well as from local islands and archipelagoes. A great thickness of material accumulated on the sea floor in Permian, Triassic, and Jurassic times.

Because New Zealand was under the sea for most of Permian time, rocks and fossils of this age are well represented in modern New Zealand. A small patch of middle Permian rocks occurs at Whangaroa Harbour, Northland. Other Permian rocks occur in broad belts extending southwest from D'Urville Island and Stephens Island through Nelson to Tophouse and Matakitaki. There Permian and other rocks are cut by the Alpine Fault and are displaced some 450 km southwest to northeast Otago, where they reappear in the Skippers Range, Hollyford Valley, and Routeburn Valley. Permian rocks then curve in a double arc through Otago and Southland. One arc traverses the Eglinton Valley, Livingstone Mountains, Takitimu Mountains, and Longwood Range and reaches the coast near Bluff. The other arc traverses the Humboldt, Ailsa, and Thomson Mountains, and the hills north of Mossburn, Lumsden, Gore, and Clinton and reaches the coast at the mouth of the Clutha River.

Three major glacial phases separated by warmer periods have been recognised in the Permian of Australia. Direct evidence for glacial activity is absent from New Zealand, as it was under the sea for most of the time, but marked phases of cooling and warming are recorded in the marine faunas. These phases are undoubtedly correlated with the waxing and waning of the nearby Australian and Antarctic ice sheets.

New Zealand Permian marine faunas include representatives of all major invertebrate groups. Brachiopods were very common, showing a great diversity of form and a range of adaptation to a variety of environments. Bivalves were also reasonably plentiful, including shells similar to modern mussels and scallops. Some of the mussel-like shells were locally so abundant that they accumulated as great shell banks (now preserved as limestone). Gastropods, bryozoa, and crinoids also inhabited the New Zealand Permian seas. Coiled cephalopods (ammonoids) were comparatively rare in New Zealand. They were more common in tropical-subtropical areas of the globe. Remains of fern-like plants occur at several Permian localities, indicating closeness to land.

The New Zealand Permian marine faunas had extremely close links with Australia and reflected similar climatic fluctuations. Other links extended to "southern" Gondwana countries. The general pattern of these "southern" links is shown (coloured pattern) on Fig. 33, and some typical "southern" fossils are illustrated (clockwise from the top): the bivalves *Myonia* and *Etheripecten*; the brachiopods *Aperispirifer*, *Capillonia* and *Terrakea*; a leaf of the seed fern *Glossopteris*. The distribution of *Glossopteris* is shown by the symbols on the map.

At intervals throughout the Permian, (presumably related to periods of mild climate in between glacial episodes), marine links extended northwards to Timor, Thailand, and the Himalaya. The Tethys seaway provided a marine corridor linking Australasia and Eurasia throughout Permian and Mesozoic time (Fig. 3).

Beginning in Permian times (and possibly earlier, although there is no geological record) New Zealand and New Caledonia had a very close relationship, often with virtually identical rocks and fossils in both countries. The closeness was such that both countries probably shared the same coastline in the Permian and, later in the Triassic, Jurassic, and Cretaceous, the same landmass. The close relationship continued until the end of the Eocene, 37 million years ago, when the South Fiji Basin and other areas of deep ocean opened up and physically separated New Zealand and New Caledonia.

Tropical reef corals and fusulinids are found at Northland's Whangaroa Harbour and Otago's Waitaki Valley. They show marked affinity with Southeast Asia, Asiatic U.S.S.R., Manchuria, Mongolia, China, and Japan, all regions that occupied positions in the central tropics in Permian times. The significance of these fossils is difficult to evaluate. Either they represent an extremely warm phase in between glacial periods, or they belong to displaced blocks of terrain rafted southwards by continental drift movements, which later became welded onto New Zealand.

Fig. 34 A reconstruction of the inland Southland region (Takitimu Mountains and Wairaki Hills) in Middle Permian times, 260 million years ago. Virtually continuous volcanic activity was occurring and the seas covering much of New Zealand at this time were dotted with arcs and archipelagoes of volcanic islands.

Although very little continuous land was probably present, it was presumably vegetated with members of the *Glossopteris* flora, that was very widespread on adjacent areas of Gondwanaland (Australia, Antarctica, India, South Africa, South America). Representatives of the *Glossopteris* vegetation are shown in the diagram

bottom right. The bottom left diagram depicts a sample of Middle Permian sea floor life, and includes the crinoid (sea-lily) *Tribrachiocrinus*, bryozoan (sea-moss) *Fenestella*, bivalves *Atomodesma*, *Etheripecten*, *Myonia*, and *Vacunella*, the gastropod *Walnicholsia*, and the brachiopods *Aperispirifer* and *Capillonia*.

In the Triassic the same global movements continued that had earlier swung the continents of eastern Gondwanaland across the South Pole (Fig. 32, 33). In the Triassic the continents gradually moved away from the South Pole. Thus as Triassic time progressed the amount of land in the polar regions steadily diminished and by the late Triassic (illustrated by the map), most of the major Gondwana landmasses were situated northwards of 60–70°S latitude.

No signs of glacial activity are known from any rocks of Triassic age, so it is probable that the ice sheets built up in the Carboniferous–Permian had largely dissipated by the start of Triassic time. Correspondingly, cold climatic zones, related to the presence of polar ice sheets, were probably also absent. Climates world-wide were correspondingly more equable than in the preceding Permian and Carboniferous periods.

Dissipation of Permian ice and warming climates of the Triassic spurred reptilian evolution and, for this reason, the Mesozoic Era is often called "The Age of Reptiles". As the ice sheets disappeared numerous reptile groups appeared in both Laurasia and Gondwanaland. At first, in the early Triassic (235–225 million years ago), Laurasia and Gondwanaland had separate and distinct reptile populations. The Laurasian reptiles gave rise to the dinosaurs, whereas those of Gondwanaland were mammal-like in

appearance, and had the same shape and size of modern wolves and foxes. However, in the middle Triassic (225–208 million years ago), differences between Gondwanaland and Laurasia became progressively blurred. This trend continued into the late Triassic (208–192 million years ago) and Jurassic (192–135 million years ago), when many land reptiles were able to roam far and wide across both super-continents.

The gradual breaking-down of distinctions between Laurasia and Gondwanaland as Triassic time progressed was probably related to the availability of suitable land links, and the increasing similarity of climates of both super-continents, as Gondwanaland moved further away from the South Pole. In late Triassic times some reptilian groups achieved wide distribution. Dinosaurs, for example, became widespread throughout Gondwanaland and Laurasia. Nonetheless, other land groups, especially plants, remained distinctively Gondwanan in aspect. These included the fern-like *Dicroidium* and many of the early conifers (see Fig. 37). The cool-temperate climate of New Zealand, the most southerly (70°–80°S latitude) of the Gondwana lands in Triassic times, may have served as enough of a deterrent to even the hardiest reptile. For this reason it is thought that New Zealand probably only became accessible to cold-blooded animals such as reptiles in

succeeding Jurassic times (see Fig. 38), when it had moved northwards into mid-latitudes, and presumably had a more favourable and warmer climate.

The map (Fig. 35) shows the distribution of two groups of land reptiles. *Lystrosaurus* (lower drawing), an early Triassic mammal-like reptile (or therapsid), is known from South Africa, Antarctica, and India. Other reptiles with whom it is closely associated are known from Australia and South America. It achieved wide distribution across Gondwanaland and (being a land dweller, with a heavy barrel-like body some 1.5 m in length, large head, stout legs, broad feet and short tail), it undoubtedly could only travel across dry land. For this reason the distribution of reptiles like *Lystrosaurus* provide crucial evidence for continuity of land in Gondwanaland in Triassic times.

Saurischian dinosaurs (upper drawing), were restricted to Laurasia early in the Triassic. However, in the late Triassic they also become widely distributed in Gondwanaland, as indicated on the map. Mammal-like therapsid reptiles, restricted to Gondwanaland in the early Triassic, migrated into Laurasia in the late Triassic. Land routes linking Laurasia and Gondwanaland, as well as their component parts, were therefore freely available in late Triassic times.

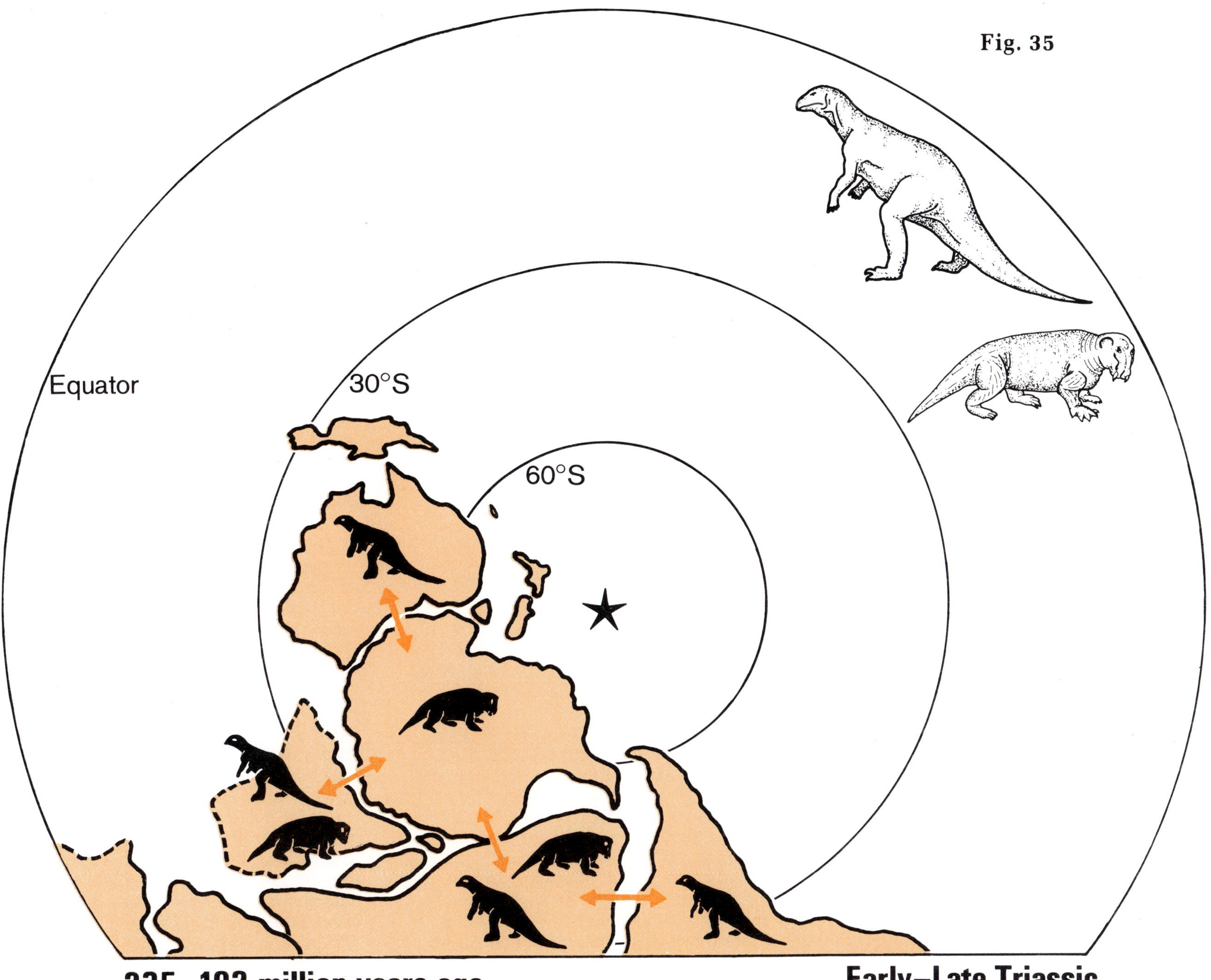
Fig. 35
Equator
30°S
60°S
235–192 million years ago
Early–Late Triassic

Slow rotation of the Gondwana landmasses, while swinging other continents away from the South Pole, had the effect of moving New Zealand even closer. Throughout most of Triassic time New Zealand was situated far to the south, at latitudes of 70°–80°S, and was much closer to the South Pole than in the Permian. However, because the world was ice free in the Triassic, New Zealand probably had cool-temperate climates throughout the Triassic, similar to those of modern Stewart Island, Campbell Island, and Auckland Islands. The comparative coolness of New Zealand Triassic climates, compared to those of the rest of Gondwanaland, is reflected by distinctive cool-temperate "Maorian" marine faunas. These faunas had links with New Caledonia, New Guinea, and via Antarctica to southern South America. The existence of such links indicates these countries were at similar southern latitudes and were joined by shallow water marine routes, along which cool-temperate organisms could migrate.

Other groups of animals and plants had a Gondwanan rather than southern aspect to their distribution. Some of the Triassic plant groups were distinctively Gondwanan. A good example was the fern-like plant *Dicroidium*, found in Triassic rocks of all Gondwana countries, as shown on Fig. 36. Gondwana forests of the Triassic also included early cone-bearers (conifers): ancestors of the modern kauri (*Agathis australis*) and other araucarian pines (e.g., the modern Norfolk pine, and the New Caledonian pine). The Gondwana forests may also have included many early ferns as well as ancestors of another group of specialised pines, the podocarps, that later gave rise to New Zealand native forest trees such as totara, rimu, miro, matai, etc.

As in the Permian, most of New Zealand continued to lie under the sea. Great thicknesses of sedimentary debris, eroded off adjacent areas of eastern Gondwanaland (eastern Australia, Tasmania, and Antarctica), together with material derived from volcanoes on land and in the sea, were deposited on the ocean floor. The record preserved in these deposits indicates that in the Triassic there was probably more land existing in the New Zealand region than in the Permian. Much of this land was of volcanic origin and would be rapidly eroded. Other land areas, formed by uplift of consolidated and hardened sediments originally deposited on the sea floor, may have been more permanent.

The new land areas created in New Zealand during the Triassic enabled land links to be established to the remainder of eastern Gondwanaland. Such links (indicated by the arrows on the map) evidently enabled *Dicroidium* to enter and colonise New Zealand. It is also thought that at least some of the ancestral stocks later to develop into the kauri, podocarps, araucarians, ferns, and lycopods (typical of much of the modern native forests of New Zealand and New Caledonia) came to New Zealand in the Triassic, probably via the same land routes as *Dicroidium*. However, New Zealand in the Triassic may have been too cool to allow entry of the large reptiles that abounded elsewhere in Gondwanaland (see Fig. 35).

TRIASSIC LIFE

At the end of the Permian a massive number of extinctions brought to a close the era of ancient life, or "Paleozoic", and ushered in the era of middle life, or "Mesozoic". Many groups died out and others were drastically thinned. Trilobites and fusulinids disappeared and brachiopods were decimated (of over 125 brachiopod genera, only 2 survived into the Triassic). Similar losses occurred in the main groups of reptiles and amphibia. However, the reptiles that survived into the Triassic expanded in numbers, varieties, and geographical ranges, to usher in "The Age of Reptiles".

Two main groups of reptiles, the rhynchosaurs and the mammal-like therapsids, dominated land life in early and middle Triassic times. As world climate became more equable (generally tropical–subtropical) and vegetation changed from lycopods and *Dicroidium* to members of early conifer families, both groups went into a decline: the mammal-like therapsids in the early Triassic, and the rhynchosaurs in the late Triassic. The decline of these groups, presumably triggered by a decline of their food plants, left the way open for the first of the dinosaurs to expand and fill the various ecological niches thus vacated. These early dinosaurs of the late Triassic were later (in the Jurassic and Cretaceous) to expand to such an extent that dinosaurs completely dominated the land life of the times.

Changes in vegetation also provided opportunities for the development of other

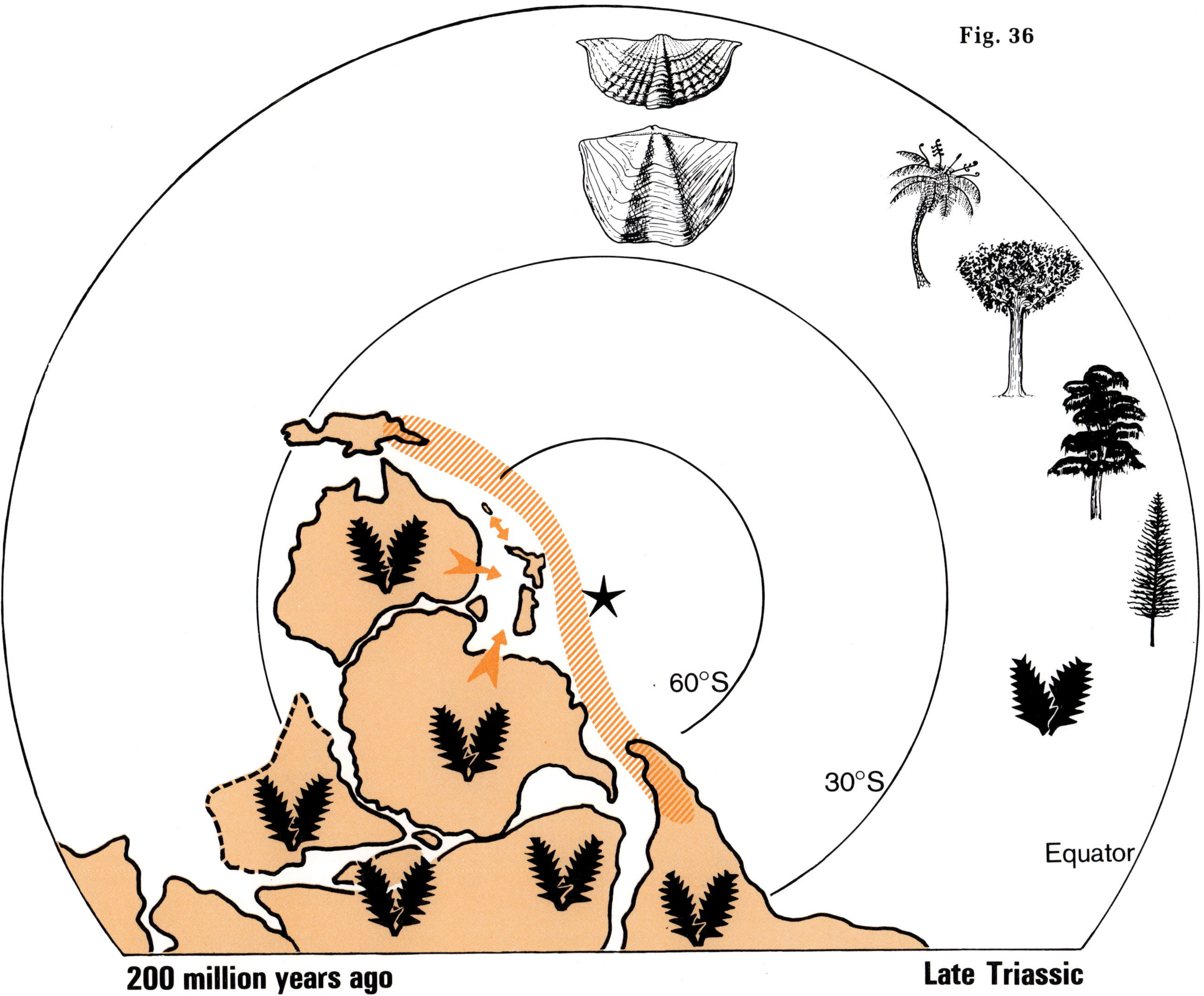

Fig. 36
60°S
30°S
Equator
200 million years ago
Late Triassic

animal groups. The first true mammals were among the new comers in late Triassic times. Mammals are characterised by the *mammae* (Latin, meaning breasts or teats) with which they suckle their young. They are also warm blooded, possess hair and have a specialised skull and jaw structure. However, the early mammals of the late Triassic were far from being the powerful predators they are today. Instead, they were tiny and obscure, about the size of modern rats and mice, and fed on insects and other small creatures.

Some scientists think the ancestors of the modern platypus and echidna of Australia (the only living monotremes or egg-laying mammals), were part of the late Triassic radiation of early mammals and mammal-like reptiles. However, whether they came to Australia at that time, or used the land routes that came into being in Jurassic and Cretaceous times is open to conjecture.

Various groups of ammonoid cephalopods and bivalves went into a decline at the end of Permian times, but, like the reptiles, revived in the Triassic to become diverse and widely distributed marine groups. Sea lilies and bryozoa were particularly abundant in Triassic seas. Among the corals, the old Paleozoic groups became extinct at the end of the Permian. By middle Triassic times, however, a new order of corals had appeared, to which belong the modern reef-building corals.

The early Triassic land floras were mainly survivors from the Permian period, consisting of clubmosses, horsetails, seed ferns, ferns, and lycopods. In the middle Triassic such groups steadily gave way to new groups of conifers, cycads, ginkgos, and true ferns— many of which survive today as "living fossils" in isolated parts of the world (notably New Zealand, New Caledonia and western China).

TRIASSIC IN NEW ZEALAND

The New Zealand fossil record includes a representative sample of Triassic animal and plant life. The overall content, however, reflects New Zealand's position in the cool-temperate zone, and coral reefs and other tropical groups are consequently absent. The lack of substantial areas of land is also reflected in the fossil record. Two Triassic plant localities are known (Waitaki River, south Canterbury and Mount Potts, inland Canterbury), both intercalated into an otherwise entirely marine sequence. The plant fossils present are fully representative of the main Triassic groups in Gondwanaland and include ferns, conifers, podocarps, and ginkgos. Although a great variety of large reptiles roamed the lands of the time, none are known from New Zealand. Remains of their marine counterparts, ichthyosaurs and plesiosaurs, are known from localities in both North and South Islands.

The Triassic marine faunas are very rich in brachiopods. Some developed bizarre wing-like shapes. Bivalves also became very abundant at various times, including scallop-like forms (*Daonella*, *Halobia*, *Monotis*) that, judging from their very wide geographic distribution, probably had very efficient means of dispersal through either mobile larvae or adults. Other Triassic marine invertebrates found in New Zealand are gastropods, bryozoa, sea eggs, and sea lilies. Cephalopods include forms ancestral to the modern *Nautilus*, conical chambered shells (*Atractites*), and a number of varieties of ammonoid (although not as plentiful as in the subtropical and tropical seas of the time).

"Exotic" fossils related to northern warm water faunas are also known from the New Zealand Triassic. Such fossils are thought to belong to displaced blocks of terrain, rafted to New Zealand during past episodes of sea floor spreading.

The New Zealand Triassic marine faunas include a marked "Maorian" element, presumably cool-temperate, developed in response to its southerly position. Such "Maorian" faunas were primarily restricted to New Zealand and New Caledonia, but other "southern" links existed, indicated on Fig. 36 by the coloured pattern. Two brachiopods with Maorian and generally southern affinities are shown (top centre): *Rastelligera* (upper) and *Clavigera* (lower).

Other drawings show (clockwise from the top) constituents of the land flora that probably came into New Zealand in Triassic time from adjacent Gondwanaland: the tree fern (e.g., *Dicksonia*), kauri (*Agathis*), podocarp (e.g., *Podocarpus*), *Araucaria*, and fronds of *Dicroidium*.

Triassic rocks are found in three main areas:
1. Twin belts which extend from Kawhia to Awakino and from Huntly to the Mokau River.
2. A belt which extends southwest from Nelson, and flanks the eastern side of the Waimea Depression.
3. A belt from the Fiordland divide and Hollyford Valley which divides into two parts through Southland. The northernmost part traverses the northern Taringatura Hills, North Range, northern Hokonui Hills, and Pukerau, and meets the Otago coast at Nugget Point. The southernmost part traverses the Wairaki Hills, southern Taringatura Hills, southern Hokonui Hills and Mataura.

Isolated pockets of Triassic rocks are also found in inland Canterbury and Otago. The rocks comprising the main ranges of both North and South Island contain Triassic fossils in several areas; especially in inland Canterbury and in the Ruahine, Tararua, and Rimutaka Ranges.

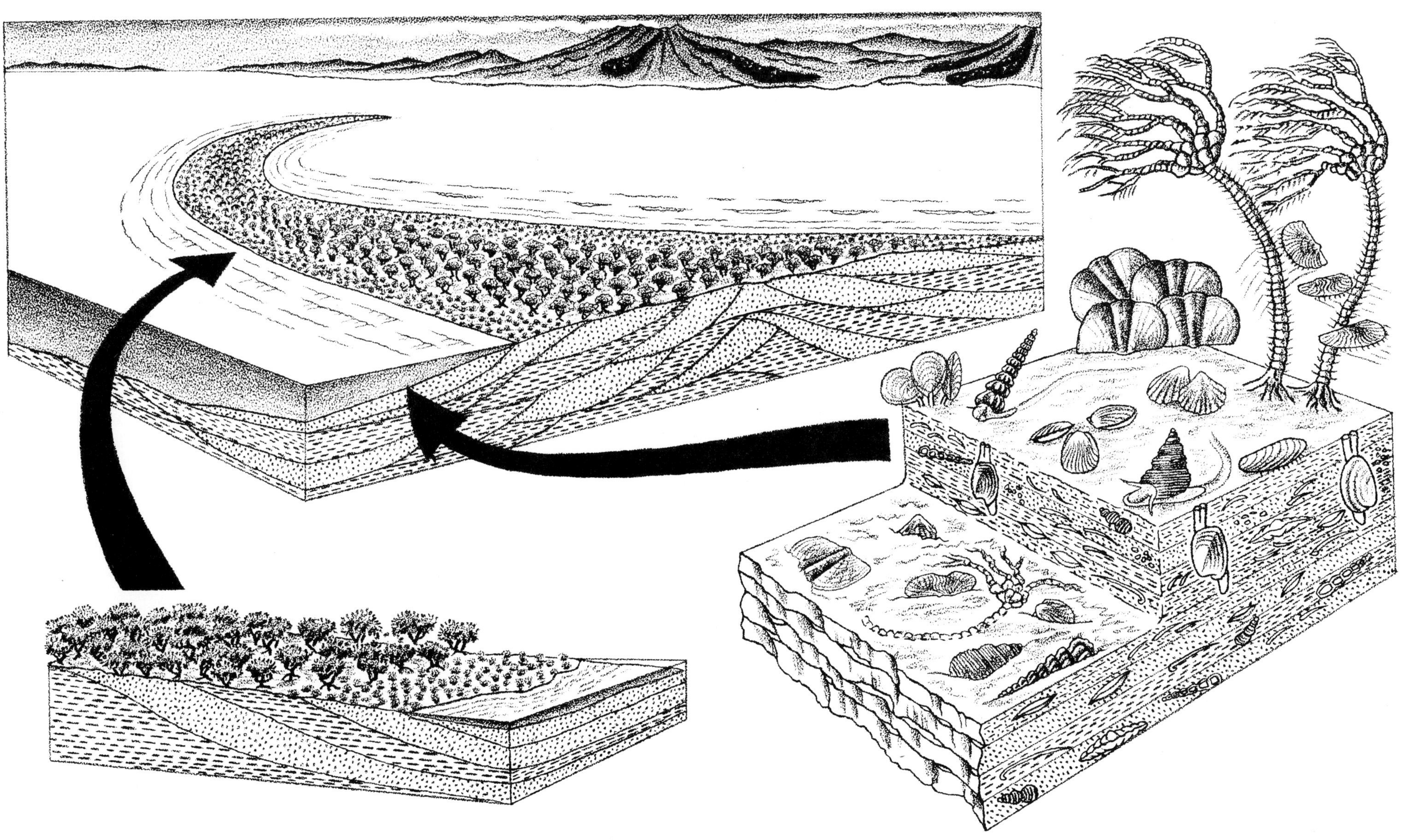

Fig. 37 A reconstruction of the Waikato coast in the middle Triassic, 225–208 million years ago. Sporadic volcanic activity was occurring on the land and much volcanic material, including ash, was being laid down in the sediments on the sea floor. Coastal areas were vegetated by *Dicroidium* and various ferns. The *Dicroidium* plants formed heath-like growths on river deltas and scrubby woodlands around the margins of coastal lagoons and swamps. A representation of the *Dicroidium* flora is shown in the diagram at bottom left. The bottom right diagram depicts a sample of Middle Triassic sea floor life and includes the crinoid (sea-lily) *Isocrinus*, the brachiopods *Spiriferina*, *Athyris* and *Rhynchonella*, the gastropods *Poroa* and *Raha*, and the bivalves *Halobia*, *Lima*, *Cucullaea*, *Anodontophora*, and *Oretia*.

EARLY & MID JURASSIC

(192–160 million years)   Fig. 38

Rotation of Gondwanaland away from the South Pole continued in the Jurassic. The effect of this rotation changed New Zealand's geographical position from 70°–80°S latitude in the Triassic to 60°–70°S in the middle Jurassic, as depicted by Fig. 38. As noted for the Triassic, the shift of land away from the polar regions substantially improved world climate and Jurassic climates were appreciably more equable than those of the present day. Temperate conditions extended into what are now the Arctic and Antarctic regions. There was a corresponding expansion of the tropical and subtropical zones, with warm climatic conditions extending north and south into areas of the globe with temperate climates today.

Probably because all the water formerly locked up in glaciers and ice sheets had flowed back into the ocean, world sea levels were comparatively high in the Jurassic. Sea levels peaked in the late Jurassic, when almost 25% of the present continental areas were covered by sea. Most of what is now western Europe was covered by shallow seas. Shallow seaways traversed the margins of Africa, Asia, India, Australia, North and South America. Most of these seas were interconnected at various times and collectively were called "Tethyan". The animals living in these Tethyan seas also extended to New Zealand, but not until the Middle Jurassic.

In the early Jurassic, New Zealand marine faunas were similar to the Maorian faunas of the Triassic, indicating that the cool-temperate conditions of the Triassic continued into the early Jurassic. However, the beginning of an influx of Tethyan groups in middle Jurassic, continuing into the late Jurassic, indicates that by those times climate had improved to warm-temperate or subtropical.

Throughout Jurassic times the earth movements that had been evident during the Triassic in New Zealand became even more marked. The movements rucked up and elevated above sea level much of the sedimentary and volcanic material deposited on the sea floor during the Permian, Triassic, and early Jurassic. As Jurassic time passed, more and more land appeared. Eventually, in the middle and late Jurassic, an extensive new landmass stretched from New Caledonia (and beyond) in the north, to Lord Howe Rise in the west, Chatham Islands in the east, and the southern edge of the Campbell Plateau in the south. Such a landmass was probably also in close physical contact with Australia and Antarctica. (The Tasman Sea and Southern Ocean were not then in existence.) Long fingers of land and interlinked chains of islands probably also extended northwards from New Caledonia towards New Guinea and Indonesia. The development of such large land areas in the middle and late Jurassic were probably the greatest extension of land ever present in the New Zealand region. This undoubtedly provided unique opportunities for the distribution of marine and terrestrial flora and fauna.

The availability of land links, combined with a climate more closely matching that of the adjoining areas of Australia and Antarctica, provided conditions for a major influx of Gondwana animals and plants into New Zealand. As in the Triassic, the vegetation of Gondwanaland was dominated by ancestors of the modern ferns, araucarians, kauri, and podocarps. Some early representatives of these groups may have colonised New Zealand in Triassic times, perhaps following the same routes from Gondwanaland used by *Dicroidium* (Fig. 35). Nonetheless, major influxes of these plant groups probably also occurred in the middle and late Jurassic, when land links were so much better, and the climate of primaeval New Zealand had improved to warm-temperate.

Wide-ranging groups of amphibia and reptiles dominated the land faunas of Gondwanaland. It is highly likely that, along with the plants just mentioned, New Zealand also received representative samples of these animals. However, changes were occurring in the land fauna. The first birds evolved from reptilian stock in the late Jurassic, but marsupials and placental mammals did not appear until Cretaceous times (compare Fig. 42, 46). Therefore, while establishment of improved land links between New Zealand and the Gondwana mainland probably provided access for amphibians and reptiles during the middle and late Jurassic, and early birds in the late Jurassic, it was still too early for the later-evolving marsupials and placental mammals.

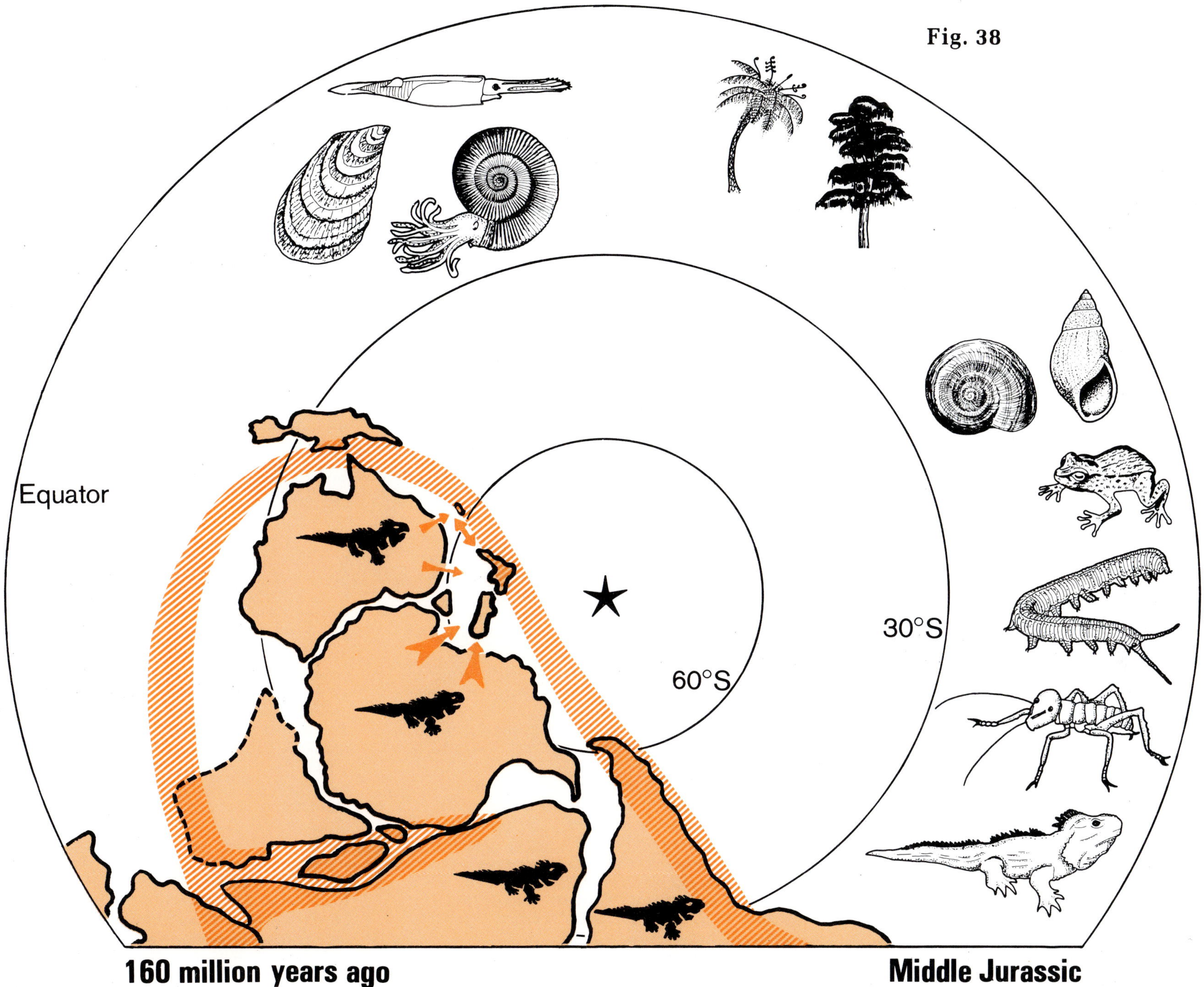

160 million years ago

Middle Jurassic

Ancestors of the native frog *Leiopelma* probably arrived in New Zealand at this time, as did those of the tuatara (*Sphenodon*). *Leiopelma* is a member of an archaic amphibian group and the tuatara is a member of a group of rhynchocephalian (meaning "beak-headed") reptiles particularly widespread in both Laurasia and Gondwanaland in Triassic and Jurassic times. (See the silhouette drawings on the map.) It is possible that early stocks of both the tuatara and *Leiopelma* may have penetrated to New Zealand in Triassic times, using routes followed by *Dicroidium* and other plants (compare Fig. 36), but this is considered unlikely as New Zealand's cool climate at that time was probably a deterrent to such cold-blooded creatures.

Other animals that probably also crossed from Gondwanaland in middle and late Jurassic times to populate the primaeval New Zealand landmass were the ancestors of *Peripatus*. They belong to a very ancient group of animals (Phylum Onychophora or "claw bearers"), with a body divided into segments like the worms, but with other features such as clawed legs possessed by modern arthropods (millipedes, centipedes, insects, etc.). They form a link between the worms and the arthropods. *Peripatus* can today be found under moist logs littering the floor of our native forest. Other animal groups that reached New Zealand at this time include the native earthworms, native snails (e.g., *Paryphanta*), some insects (notably the weta, the giraffe weevil *Lasiorhynchus barbicornis*, and some spiders), native slugs, freshwater crayfish, freshwater mussels, and some freshwater fish. All of these land organisms were probably originally widely distributed across eastern Gondwanaland and their probable routes into New Zealand and New Caledonia are indicated by arrows (large =

routes of probable major importance; small = minor importance).

As well as providing migration opportunities for land organisms, development of the primaeval New Zealand landmass also opened up shallow-water marine routes to the north (Indo-Pacific, Southeast Asia) and to the south (Antarctica, South America). Shallow-water marine organisms became widely distributed around the margins of Gondwanaland (as indicated by the diagonal pattern on the map). The influx of terrestrial

creatures into New Zealand in the middle and late Jurassic was matched by major incomings of shallow-water marine animals.

Illustrations on Fig. 38 represent middle Jurassic immigrant groups. They are (clockwise from top left): the bivalve *Inoceramus*; the squid-like belemnite *Hibolithes*; the ammonite *Dactylioceras*; the forest plants tree fern and podocarp; the land snails *Paryphanta* and *Placostylus*; the native frog *Leiopelma*; *Peripatus*; the weta *Deinacrida*; the tuatara *Sphenodon*.

MIDDLE JURASSIC IN NEW ZEALAND

As more and more land was created so the sea was progressively pushed away from the area now occupied by continental New Zealand. For this reason the record in New Zealand of middle Jurassic marine rocks is rather fragmented. Middle Jurassic rocks occur in two main areas: the Waikato coast and inland Southland/Otago. In both areas the rock sequences show many signs that they were deposited close to land and are interrupted at intervals by non-marine beds, often rich in plant material. The sequences in Southland and Otago do not extend beyond the middle Jurassic, presumably because after that time the entire area became land. The Waikato coast marine sequences, on the other hand, continue into the late Jurassic. As with the Triassic, patches of middle and late Jurassic rocks occur elsewhere in New Zealand, associated with the main axial ranges, in inland Canterbury, Marlborough, and the southern North Island.

The progressive shallowing of the sea that occurred throughout middle Jurassic time is reflected in the abundance of plant remains and fossil forests, especially in the South Island, where land influences at this time were more apparent than in the North Island. Beds of fossil plants are known from Waikawa and Curio Bay (southeast Southland), Owaka (Catlins Coast, southeast Otago), Hokonui Hills, Mataura Falls and Mokoia (inland Southland), Malvern Hills and Benmore (inland Canterbury). Te Maika,

Kawhia Harbour is the most notable North Island locality with plant remains and fossil trees of middle Jurassic age.

The fossilised trees of an extensive fossil forest are exposed on the shore platform and in sea cliffs at Curio Bay, Southland. The trees are rooted in green sandstones and blue shaly clays containing plant impressions. Silica minerals have replaced the entire woody structure of the trees and rendered them extremely resistant to erosion. Thus they withstand the action of the sea much longer than the surrounding rocks and are exposed in relief by the erosion. In the past, erect tree trunks have been standing in the face of the sea cliffs, with their entire root systems exposed in the underlying beds. The sandstones are almost flat lying and at low tide form wide shelving ledges over which are strewn many tree stumps and prostrate trunks. On occasion trunks over 15 m in length have been measured and some have exceeded 30 m. Fossil wood has been obtained at intervals along the Waikawa coast for some 13 km and also inland near Waimahaka, indicating that the fossil-forest beds were probably of considerable extent. Some of the fossil trees are related to the modern kauri and Norfolk Island pine *Araucaria*, Most of the fossil tree trunks show very well developed seasonal growth rings which indicate that at the time New Zealand had a climate with well-defined seasons.

Fig. 39 Bones of fish-lizards (ichthyosaurs) occur in late Triassic and Jurassic marine rocks of New Zealand. Although reptilian, they had fish-like bodies and in this respect are aptly named. The entire body was modified for life in the sea and was superficially like a dolphin—a mammal similarly adapted to marine life. The average length of ichthyosaurs was about 2.5–3.0 m although some attained lengths of 13 m. The skull was armed with numerous large and sharp teeth, sometimes as many as 200. The limbs were converted into fin-like flippers. The strong fish-shaped body had a large leathery vertical dorsal fin and a crescent-shaped tail fin. The ichthyosaurs preyed mainly on fish, crustacea (crayfish, etc.), and cephalopods (ammonites and belemnites). However they also shared the same seas with other equally rapacious reptiles, notably the plesiosaurs (Fig. 46) and, judging from traces of severe injuries seen in some ichthyosaurs, they in turn were probably preyed upon. Injuries may have also been sustained during fighting between rival ichthyosaurs at mating time, as depicted in this drawing.

Continued rotation of Gondwanaland away from the South Pole swung New Zealand even further northwards from its middle Jurassic position. In the late Jurassic, depicted on the map, New Zealand was lying at about latitude 55°S and the subtropical/warm-temperate conditions that were first evident in the middle Jurassic became even more marked.

Evolution of reptilian stocks proceeded apace in the Triassic and Jurassic and as well as dominating the land, some reptiles entered the sea and developed paddle-like limbs and a streamlined shape. They gave rise to the dolphin-like ichthyosaurs and the long-necked plesiosaurs. Other reptiles extended large sail-like flaps of skin between their sides and greatly elongated fingers. Although some of these creatures, called pterosaurs, attained wing spans of up to 15 m it is doubtful they could actually fly by flapping their wings. Rather, they were probably gliders. Some reptiles, however, formed feathers from their reptilian scales and developed light, hollow bones and a large keel-like chest bone to provide attachment for powerful wing muscles. These feathered reptiles were the ancestors of modern birds. With scale-clad legs (still present in modern birds: a hint of their reptilian ancestory), reptilian teeth, and clawed wings (both lost in modern birds), they first appeared in the middle of the late Jurassic, about 150 million years ago.

As evolutionary changes proceeded within the early birds other groups appeared, less like reptiles, but nonetheless with many primitive features, compared with more advanced birds. A notable early group of birds was the Ratites, ancestors of the moa and kiwi (New Zealand), ostrich (Africa, and in the past Europe and Asia), rhea (southern South America), emu (Australia), cassowary (New Guinea and Australia), and the extinct elephant bird (Madagascar). Except for the kiwi, all these birds are large and flightless and look very much alike.

The significance of the flightless Ratites and their relationships to other flying birds has been the subject of much speculation. It is now recognised that Ratites have arisen from flying ancestors by degeneration of wings and breast muscles as they increased in size and began to depend on their running ability to escape from enemies. It has also been accepted that ancestral Ratites evolved large body size and became flightless soon after their initial development, and then were distributed across the southern continents when Gondwanaland was in existence and when continuous land links were available. Although some living Ratites can swim for very short distances they cannot cross ocean barriers. Everything points to the probability that the ancestors of the moa and kiwi walked into New Zealand before it became isolated from the other Gondwana continents.

It is not clear when the Ratites appeared as a separate group of birds. It is thought that the Ratite ancestors originated in South America. The tinamous, a family of grouse-like birds living today in some parts of South America, are presumed to have split off from the main Ratite group at about this time. The ancestral Ratites subsequently lost the power of flight and Ratite groups soon became fairly widely distributed across Gondwanaland, as suggested by the silhouette drawings on the map. One early group made its way from South America across West Antarctica and into New Zealand, there to eventually give rise to the moa and kiwi. Although it is not known when this group made its migration, it is likely that it was probably not later than latest Jurassic or earliest Cretaceous. Land links between New Zealand and other Gondwana continents were probably at their maximum extent at that time. Although southern land links persisted into the middle Cretaceous (compare Fig. 42, 44), they were probably more restrictive in terms of climate and accessibility.

Land routes into New Zealand and New Caledonia that were probably available to the ancestral Ratites in late Jurassic times are indicated by the arrows on the map (large arrows = probable major routes; small arrows = minor routes).

JURASSIC LIFE

As well as the development of the birds from reptilian groups, the Jurassic period also saw establishment of ancestral stocks that later in the Cretaceous gave rise to the mammals. Mammal-like reptiles (therapsids) had co-existed with the dinosaurs throughout Triassic time and by the late Triassic had become almost as common and as widespread as the dinosaurs. At the end of the Triassic, however, they went into a

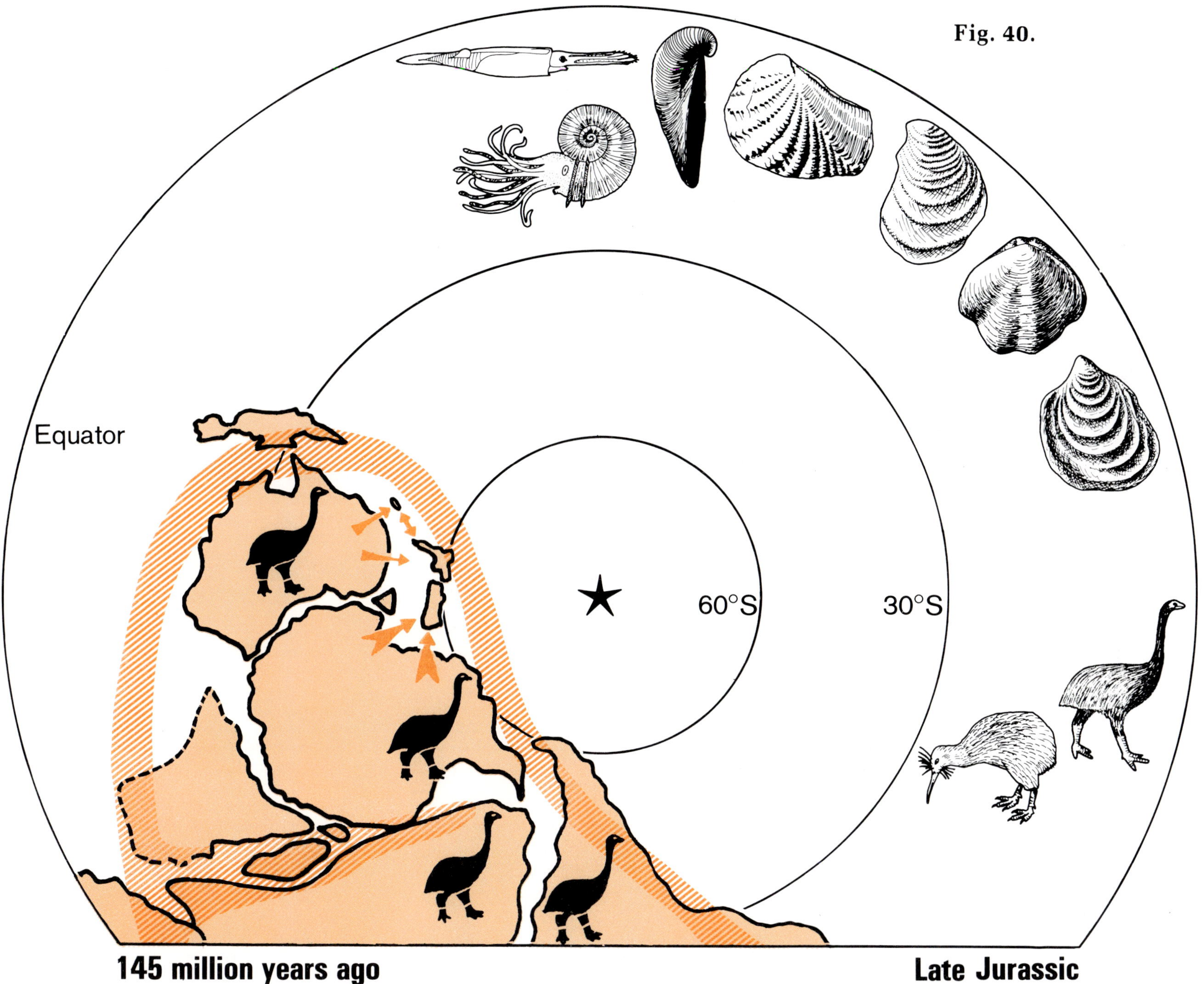

Fig. 40.
Equator
60°S
30°S
145 million years ago
Late Jurassic
59

major decline and only a few groups continued into the Jurassic. Then throughout the Jurassic they were overshadowed by the dinosaurs and, by comparison, were markedly insignificant both numerically and in size (they were about the size of a mouse or rat).

Although ichthyosaurs and plesiosaurs tended to dominate the vertebrate marine fauna, turtles, and crocodiles were also common in tropical seas. Modern fish groups appeared towards the end of the Jurassic. In many parts of the world the Jurassic invertebrate marine faunas were dominated by

ammonites. Although only one ammonite family survived the rash of extinctions at the close of the Triassic, later during the Jurassic it radiated to produce a great diversity of forms, many achieving worldwide distribution. Belemnites, squid-like relatives of the ammonites, also came into prominence during the Jurassic.

Bivalves, gastropods, and crustacea flourished. Brachiopods were more abundant and diverse than they are today. Sea lilies that had been prominent members of sea floor life in earlier times, declined in importance, giving way to ancestral groups of modern

starfish, sand dollars, and sea urchins. Reef-building corals related to modern forms were common in warmer seas and calcareous algae and sponges locally formed reef-like masses. The marine plankton was well represented and included foraminifera, ostracods, radiolaria, and coccoliths.

The Jurassic land flora was dominated by non-flowering plants (gymnosperms), forests of which covered much of the terrain. They included conifers, gingkos, and cycads. Ferns and horsetails made up the remainder of the land flora.

LATE JURASSIC IN NEW ZEALAND

Steady uplift of land throughout the Jurassic pushed the sea away from many of the areas now occupied by land in New Zealand. Judging from the presence of thick beds of gravel, sand, and plant remains (indicating closeness of land), most of the South Island probably rose above the sea towards the close of the middle Jurassic. Although sea had also retreated from large areas of the North Island, marine sediments were still being deposited along the Waikato coastline and the south Auckland area. The rocks deposited along the Waikato coast were deposited very close to the shoreline, whereas those further east, in the areas now occupied by the Hunua Ranges and Coromandel, were deposited in deeper water, at the edge of the continental shelf.

The rocks deposited along the Waikato coast (between Awakino and Port Waikato), are often very rich in fossils. The fossils are a good sample of Jurassic sea-floor life. Bivalves are particularly common, including large ribbed clams (*Retroceramus*) and oyster-like shellfish (*Buchia, Malayomaorica*). Gastropods, less common than bivalves, also occur, as well as tusk shells

(scaphopods), sea lilies, starfish, crayfish, and worms. Although ammonites abounded in the tropical and sub-tropical seas of the time (then covering large areas of southern Europe and the Mediterranean) they were less abundant in New Zealand seas. Presumably this was because New Zealand had a somewhat cooler warm-temperate or marginal sub-tropical climate. For the same reason, reef corals, and many tropical varieties of bryozoa, algae, and bivalvia are absent from New Zealand. Belemnites, the squid-like relatives of the ammonites, were reasonably common in New Zealand, presumably because of their greater temperature tolerance. Sometimes layers packed with belemnites are preserved in the rocks, probably representing strandings of swarms of these squid-like animals. Some of the belemnites are up to 114 mm long, representing squids about 1.2 m long. Numbers of ammonites are also occasionally found grouped together in layers, again probably a result of stranding. Some ammonites preserve features of the original mother-of-pearl shell in exquisite detail. While most of the late Jurassic ammonites of New Zealand are between 50–200 mm in diameter, some giant forms, up to 1.52 m diameter, have been found at Kawhia Harbour and Taharoa. Although brachiopods were present in the Jurassic seas of New Zealand, the large bizarre endemic types typical of the Triassic were absent and like the ammonites they were not as

common as in some of the tropical areas. Of the larger animals, fish scales and bones of marine reptiles have been found.

As earth movements continued, and more and more land was pushed up out of the sea, the marine sequences in the Waikato and south Auckland gave way in the latest Jurassic to terrestrial beds, often crammed with fossil plant material. A notable locality for latest Jurassic fossil plants occurs at Huriwai Stream, on the coastline south of Port Waikato. This locality contains beautifully preserved ferns and horsetails a well as leaves, cones, and stems of pine-like trees. The plants lived on a large river delta, built out into the head of a bay. The ferns colonised abandoned river channels, swamps, and stream banks. Coniferous forests grew on the more stable areas, although they were often flooded by sediment from adjacent overflowing river channels.

The influx of animal groups (both land and marine), that commenced in the middle Jurassic continued in the late Jurassic. Illustrations (Fig. 40) of representative late Jurassic immigrant groups are (clockwise from the top): the squid-like belemnite *Belemnopsis*; the ammonite *Lytoceras*; the bivalves *Buchia, Myophorella*, and *Retroceramus haasti*; the brachiopod *Kutchithyris*; the bivalve *Retroceramus everesti*; the Ratite birds *Dinornis* (moa) and *Apteryx* (kiwi).

Fig. 41 A reconstruction of sea-floor life in New Zealand, 145–65 million years ago. Ammonites were very common in the seas of the time, sometimes becoming almost as common as the shellfish on modern beaches. Although most of the ammonites ranged from 75–100 mm in diameter, some attained a very large size: up to 1.5 m in diameter. The sea-serpent-like mosasaurs attained lengths of 14–15 m and judging from their stomach contents and injuries inflicted on ammonites, they enjoyed an ammonite meal. The dolphin-like ichthyosaurs ranged in length from 2.5–10 m. Squid-like belemnites were also common at the time and ranged from 1–2 m in overall length, although giants attained lengths of 4.6 m.

The Cretaceous was a time of change. The latest Jurassic and earliest Cretaceous saw the onset of rifting movements that were to split off India and Antarctica from Africa, and separate Africa from South America. Similar splitting movements were also occurring in the Northern Hemisphere, notably along the site of the modern Atlantic Ocean. These movements marked the beginning of the division of Laurasia and Gondwanaland into their individual continents— a process still continuing today.

At about this time, the first signs of movements that were later to split New Zealand away from Australia began on the site of the modern Tasman Sea. Long split-like rift valleys developed along the western margin of New Zealand, along the Lord Howe Rise and eastern margin of Australia (at this time all in close contact). The sea eventually extended into these valleys as they deepened. The general trend of the rifting movements is indicated by the saw-tooth pattern on the map.

Increasing separation of Antarctica and India from Africa began to form the Indian Ocean, and separation of Africa from South America, to form the South Atlantic Ocean. All these openings had the effect of reversing the movement of eastern Gondwanaland away from the South Pole (seen earlier in the Triassic and Jurassic , compare Fig. 36, 38, 40). As a result, New Zealand, New Caledonia, Australia, Antarctica, and southern South America were swung closer to the South Pole. The geographic position of New Zealand changed, therefore, from 55°S latitude in

late Jurassic times to 70–80°S in the early Cretaceous. Consequently, the climate of New Zealand gradually changed from warm-temperate (and perhaps marginally sub-tropical) in the Jurassic to cool-temperate in the Cretaceous. As there were no major landmasses in the vicinity of the South Pole there was probably no polar ice and correspondingly, no frigid polar climatic zones. (There is no record of early Cretaceous glacial activity.) Instead, as in the Triassic and Jurassic, the widths of the other climatic zones were increased and cool-temperate and cold-temperate zones occupied the polar areas. Therefore although New Zealand was close to the South Pole (and came even closer in middle Cretaceous times; see Fig. 44) its climate was cool-temperate in early Cretaceous, ranging to possibly cold-temperate in middle Cretaceous, and in this respect was very similar to Triassic climates.

Climatic change, from warm-temperate in the Jurassic to cool-temperate in the Cretaceous, also had the effect of discouraging the movement of organisms into New Zealand from northern Indo-Pacific sources. Many of the potential migrants from the north were probably sub-tropical or warm-temperate in character. Even if migration routes from the north existed into New Zealand at this time they were probably only available for cool- or cold-temperate organisms.

Rifting movements preceding the opening of the Tasman Sea and Southwest Pacific Ocean also disrupted the land and shallow-water marine links that had been evident in

middle and late Jurassic times, particularly those to the north and west.

The extensive landmass created in the New Zealand area in middle and late Jurassic and earliest Cretaceous time was rapidly cut into by rivers and the sea. By the end of early Cretaceous time, 110 million years ago, the land in many areas had either been worn down to a low level or flooded by the sea. This also caused steady deterioration of the land links that earlier had facilitated movement to New Zealand of ancestors of organisms such as the tuatara, moa, and kiwi.

The two factors, climatic change and disruption of migration routes from the north, were together responsible for a marked decrease in Indo-Pacific immigrants in New Zealand marine groups in the early Cretaceous. At the same time, distinctive "southern" cool-temperate shallow-water marine groups began to appear in New Zealand, related to similar groups in Australia, southern South America, Antarctica, and South Africa (indicated by the striped pattern on the map). "Southern" links were also indicated by the land organisms that probably arrived during the early and middle Cretaceous (compare Fig. 42, 44).

It is believed that ancestors of the modern protea group of plants (Proteaceae) came to New Zealand from South America in the early Cretaceous, using a southern route via western Antarctica (indicated by the silhouette drawings and large arrow on Fig. 42). Proteaceae probably also migrated

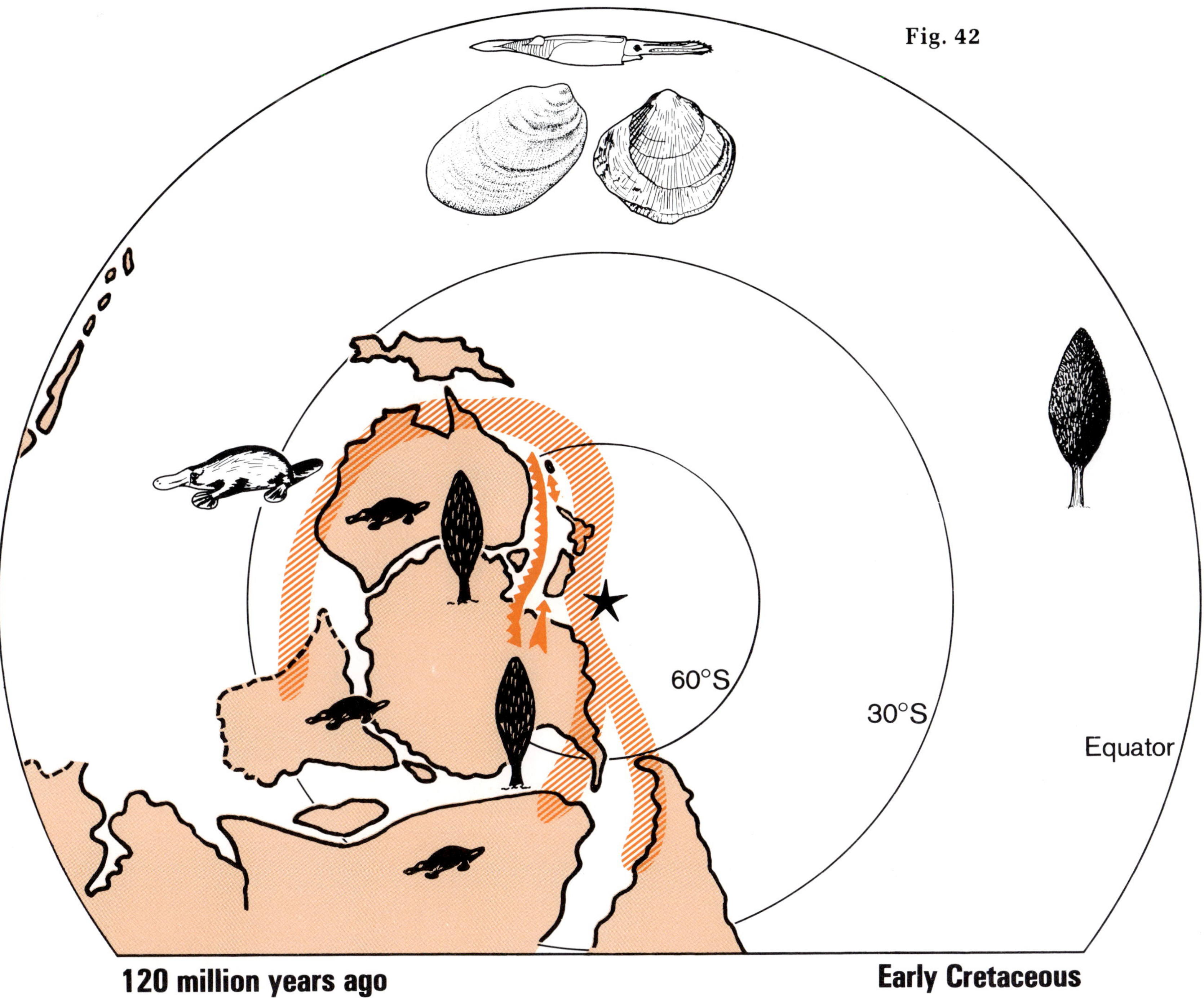

120 million years ago   **Early Cretaceous**

into Australia and South Africa at the same time. On those continents they produced the magnificent variety of proteas seen today. In New Zealand, by contrast, probably as a result of decimation during the ice ages, the protea group is today represented by only two rather inconspicuous plants: rewarewa (New Zealand honeysuckle *Knightia excelsa*) and toru (*Toronia toru*).

At about the same time in the early Cretaceous as the ancestors of the protea group were moving via southern routes into New Zealand and Australia, ancestors of the platypus and echidna may have been moving across some of the southern lands, to gain entry to Australia. These animals, comprising the only living monotremes, or egg-laying mammals, are today known only from Australia and New Guinea. Like other elements of Australia's unique flora and fauna, they were left isolated when it became separated from other continents.

Ancestral proteas were apparently able to enter both Australia and New Zealand, but it is likely that ancestral monotremes were only able to gain entry to Australia. Fundamental differences may have existed in the routes followed by the proteas, compared with the monotremes. One possibility is that the route providing access to New Zealand, and followed by the proteas, was a more southerly route, favouring organisms with cool- or cold-temperate requirements. The early monotremes, however, may have used eastern Antarctica as a stepping stone, and thus had a warm-temperate or sub-tropical route to gain access to Australia (see Fig. 42). Such a route is suggested by the silhouette drawings of a platypus on the map.

CRETACEOUS LIFE

The land floras of the earliest Cretaceous, like those of the Jurassic, were dominated by non-flowering plants (gymnosperms): conifers, ginkgos, and cycads. Early in the Middle Cretaceous, however, the first flowering plants (angiosperms) appeared. This was the most significant evolutionary step in the plant kingdom because it opened the way for plants to fully colonise the land in the Devonian. The flower has a variety of mechanisms to ensure easy pollination (by insects or by the wind) and produces resistant seeds, capable of wide dispersion. The superior method of reproduction and the variety of mechanisms for seed dispersal ensured the success of the flowering plants, so much so that in time they supplanted the earlier-evolved gymnosperm types over large areas of the world.

Most elements of Jurassic animal life continued virtually unchanged into the Cretaceous. The marine plankton included radiolarians and foraminifera. Both groups became particularly abundant towards the end of the Cretaceous. Microscopic calcareous marine algae (coccoliths) first appeared in the Jurassic. They became very abundant in the Cretaceous, especially in the middle Cretaceous, when extensive deposits of chalk, composed almost entirely of the remains of coccoliths, were built up in many parts of Europe. Today they form features such as the White Cliffs of Dover.

Sponges, brachiopods, and corals were abundant in the warmer regions of Cretaceous seas. Bivalves and gastropods flourished in shallow seas, particularly towards the close of the Cretaceous. In many regions the place of brachiopods in

EARLY CRETACEOUS IN NEW ZEALAND

Because the primaeval New Zealand landmass formed in middle and late Jurassic times was still very extensive and largely intact in the early Cretaceous (and in some areas was probably still undergoing uplift) this time period is virtually unrepresented in New Zealand. The first 23 mllion years of early Cretaceous time is apparently missing completely. Only some 2 million years at the very end of the early Cretaceous is represented by fossils, and even then the record is very meagre. Rocks of this age occur at Koranga, Raukumara Peninsula, and at the head of the Waimana and Waiotahi valleys, inland from Opotiki, Bay of Plenty. These rocks consist of very coarse materials (pebbles, rock chips, gritty sand, etc.). This indicates they were deposited very close to a rugged coastline, at the edge of the primaeval landmass (Fig. 43). The fossils include 32 varieties of bivalves, some of them resembling modern oysters, scallops, and mussels. Other animal groups are rare, but include gastropods, belemnites, brachiopods, sea-eggs, bryozoa, and worm tubes.

The squid-like belemnites and many of the bivalves and gastropods found in New Zealand early Cretaceous rocks all have close relatives in Australia, South America, South Africa, and West Antarctica. These close affinities give some idea of the way southern countries were linked by shallow water (and probably also land) routes beginning in the early Cretaceous, but continuing throughout Cretaceous time.

Some of the immigrants that came to New Zealand in early Cretaceous time are illustrated around the upper margin of Fig. 42. The illustrations depict the belemnite *Dimitobelus* (top); followed by the bivalve *Aucellina* (top left); the bivalve *Maccoyella* (top right); and a representation of the proteacean tree, *Knightia excelsa* (rewarewa, far right).

the environment was being steadily taken over by bivalves, which had the advantage of being able to burrow into the sediment, both to feed on the contained organic material and to hide from predators.

Ammonites and belemnites were conspicuous members of marine communities. Echinoderms (starfish, sea eggs, etc.) were reasonably common and resembled modern types. Fish faunas included varied and diverse shark groups. Modern bony fish underwent a major radiation during the period. Many groups of crustacea appeared in very much their modern form.

Reptiles were without a doubt the most spectacular Cretaceous animals, both on land and in the sea. Marine forms included long-necked plesiosaurs up to 16 m long. Although dolphin-like ichthyosaurs continued from the Jurassic into the Cretaceous, they were less common than in Jurassic seas. A new group of sea serpents, the mosasaurs, appeared in the Cretaceous. With massive skulls (up to 1.5 m long), and an impressive array of large sharp teeth, they were undoubtedly formidable predators. The body lengths of mosasaurs reached 14–15 m.

Equally formidable reptiles also roamed the land. Examples are the armour-plated *Ankylosaurus*, the horned *Triceratops*, and *Tyrannosaurus*, the largest known terrestrial carnivore, standing at a height of 15 m.

Fig. 43 The ancestral New Zealand landmass, created by earth movements in Jurassic time, probably occupied an area spanning the entire length and breadth of modern New Zealand as well as extending northwards towards New Caledonia, southwards to Campbell and Auckland Islands, westwards towards the Lord Howe Rise, and eastwards to beyond the Chatham Islands. The landmass probably attained its maximum geographic extent in latest Jurassic and early Cretaceous time, because this time interval corresponds to a gap in the marine geological record in New Zealand. (This indicates that land probably occupied all of the area of modern New Zealand and as a consequence no marine sediments were deposited.) As time passed in the early Cretaceous the ancestral landmass was steadily reduced in height and extent by the combined effects of erosion by rivers and streams and the sea. By 110–115 million years ago, the land had been worn down to such an extent that sea was able to flood in along the eastern areas of both the North Island and Marlborough and across Northland. Rifting was occurring along the West Coast, heralding formation of the Tasman Sea.

Land is indicated by the coloured area, the Alpine Fault by a broken line, volcanoes by cone symbols, and coal swamps by a swamp symbol. The outline of modern New Zealand is shown for reference, but it must be remembered that at this time it did not physically exist.

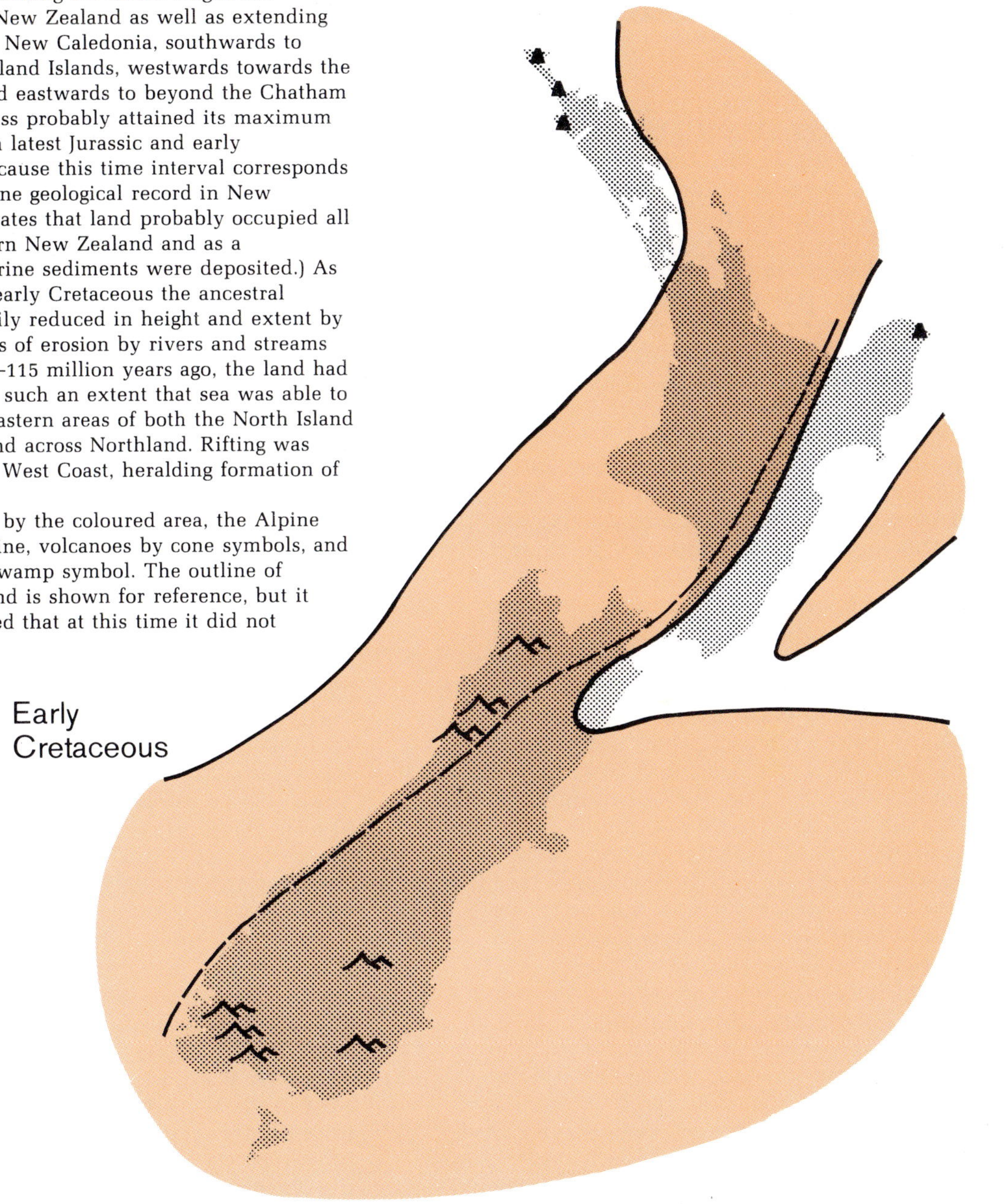

Continuation of opening movements in the Indian and South Atlantic Oceans, and further rifting in the Tasman Sea moved Australia, New Zealand, New Caledonia, and Antarctica even closer to the South Pole. Thus in middle Cretaceous times New Zealand came to lie within 5–10° of the South Pole. However, because there was no polar ice, the climate of New Zealand was not necessarily frigid, but rather cool- or cold-temperate, similar to that of modern Stewart, Campbell, and Auckland Islands.

As rifting movements intensified in the area between Australia and New Zealand, long fingers of sea extended along what was later to become the Tasman Sea. The chances of land links to the west and north became even more reduced. Erosion of the New Zealand landmass was also proceeding, and many areas around its edges were beginning to be eaten away and submerged by the sea. Movement of New Zealand close to the South Pole widened climatic differences between New Zealand and Indo-Pacific regions to the north. Thus immigration of marine and terrestrial organisms from such sources probably became very much restricted. There was little likelihood of trans-Tasman land routes and only those marine creatures with free-floating or swimming stages in their life cycle and broad temperature tolerance could move southwards into New Zealand.

Land routes persisted south from New Zealand across the Campbell Plateau and West Antarctica. Ancestors of the southern beech (*Nothofagus*) were able to enter New Zealand from this direction some time in the middle Cretaceous. Like the proteas in the early Cretaceous, *Nothofagus* first appeared in South America. Also, like the proteas, they probably used West Antarctica as a stepping stone to enter New Zealand and New Caledonia (and eventually Australia and New Guinea). Although it is thought that the first proteas entered the Southwest Pacific in the early Cretaceous, it is possible that some later immigration of the protea family also occurred in the middle Cretaceous, perhaps with *Nothofagus*. The possibility of such land routes is indicated by arrows and silhouette drawings on Fig. 44.

The southern land routes evident in the mid Cretaceous were accompanied by excellent shallow-water marine routes, as indicated by the striped pattern on Fig. 44. Elements of this southern marine fauna were also shared with Australia and India. Other marine organisms (mainly ammonites and various shellfish), capable of crossing stretches of open ocean by floating or swimming, either as larvae or adults, were able to migrate southwards into the New Zealand region from the ancient Tethys Ocean covering the Middle East and Southeast Asia.

Elsewhere in the world the middle Cretaceous was a time when the patterns of land and sea and mountain ranges as we know them today began to be roughed out for the very first time. Major movements of the continents were underway, and as a result oceanic areas were either actively widening (the Atlantic and Indian Oceans), or, as in the Tasman Sea, going through a prelim-inary cracking and splitting phase (rifting), before pulling apart.

Opening-out of oceanic areas in the middle Cretaceous was accompanied by widespread flooding of a series of continental areas. This flooding resulted from displacement of sea water by changes in the shape and volume of the ocean basins, as the continents moved apart. During these periods of flooding of the continents, oceanic waters spread over large parts of North America, Africa, and Europe. A broad seaway, up to 620 km wide, traversed the length of the western interior of North America, extending from Alaska and Arctic Canada southwards some 2000 km to the Gulf of Mexico.

In Africa sea covered North Africa and then moved across the Sahara, to link up with sea coming in from the South Atlantic across Nigeria and Chad. Sea extended over much of western Europe and some 45% of the present land area of U.S.S.R. was submerged. In Europe the widespread mid Cretaceous drowning of the land came at a time when much of the land was low. This meant that little sand, mud, and clay came from the land into the sea, and clear water conditions often prevailed. Clarity of the water allowed sunlight to penetrate and this, combined with the warm conditions, stimulated an enormous growth of microscopic calcareous marine algae (coccoliths). When the algae die and fall to the sea floor, they form a deposit called chalk. During the times of high sea level in the middle Cretaceous chalk was deposited over much of Europe, extending from Ireland to the Caspian Sea.

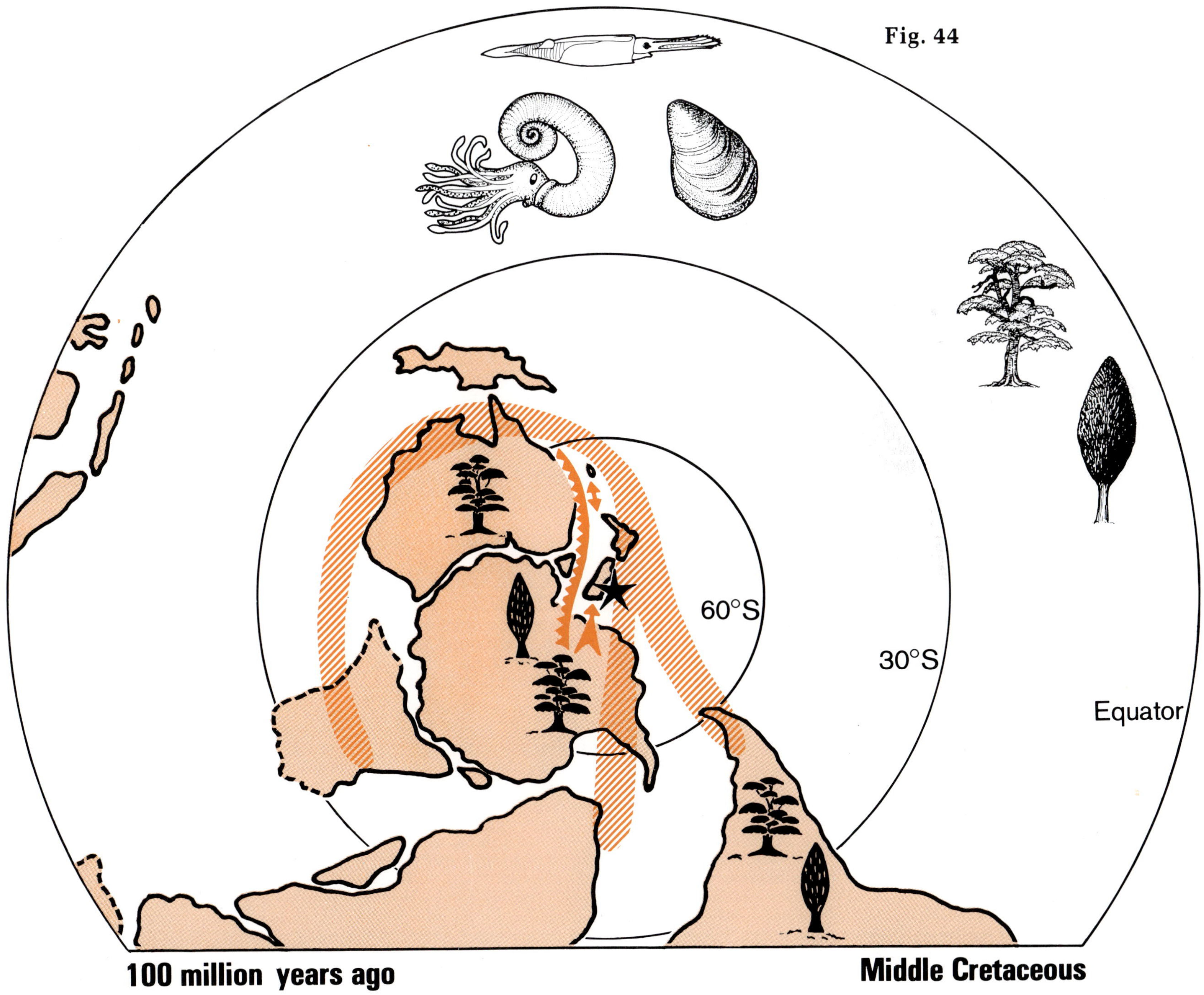

Fig. 44
60°S
30°S
Equator
100 million years ago
Middle Cretaceous

MIDDLE CRETACEOUS LIFE

The movements of continents and opening up of oceans that occurred in the middle Cretaceous had far-reaching effects on many groups of organisms. The opening up of seaways linked cold-water Arctic and north Russian seas to temperate and warm waters of the North Atlantic and the ancient Tethys Ocean. This facilitated extensive exchanges of marine faunas. Widespread transgressions of sea across the land accentuated warm maritime climates in North America, Africa, Europe, and Eurasia. Ammonites, echinoderms, and a great variety of bivalves were particularly abundant in many of the shallow seas. On the land, vegetation types and patterns similar to those we know today became established, as the dominance of non-flowering plants (gymnosperms) such as pines, firs, cedars, cycads, etc., gave way to a preponderance of flowering plants (angiosperms).

Changes also occurred in the animal life on land. About 100 million years ago, new groups of carnivorous dinosaurs appeared that were lightly built (like an ostrich) and could presumably run at a fair turn of speed. They had enormous eyes and comparatively large brains. The part of the brain concerned with muscular co-ordination was particularly prominent, indicating a highly developed sense of balance. They also had long thin fingers which could deftly snatch and grasp prey. It is likely that such dinosaurs were capable of hunting in fading light and presumably their prey included small furry mammals which came out to feed as night fell.

MIDDLE CRETACEOUS IN NEW ZEALAND

The primaeval New Zealand landmass, that came into being towards the end of Jurassic times and in the earliest Cretaceous, still occupied a large area. It was still physically intact in many parts of New Zealand in the middle Cretaceous. Consequently, much of the area that is now modern New Zealand was above sea level and the marine record of middle Cretaceous times is rather patchy. Marine rocks of middle Cretaceous age are only found in areas around the edge of the modern New Zealand landmass. Here the combined effects of marine and terrestrial erosion had worn down the old land to such an extent that the sea was able to flood in across the remnants of the eroded rocks (Fig. 43). Local areas of subsidence, related to the splitting movements then going on around New Zealand, preceding formation of the Tasman Sea and Southwest Pacific Ocean, allowed fingers of sea to extend some distance inland in a few areas.

Marine sediments of middle Cretaceous age occur in Northland, north from Kaipara Harbour. A major belt of such rocks forms the backbone of the Raukumara Peninsula and similar rocks form the coastal hills of the Wairarapa. Middle Cretaceous marine rocks extend inland along the Awatere and Clarence valleys in northern Marlborough, probably related to long inlets of the sea penetrating into the land. All of the marine rocks show signs of having been deposited close to a coastline fringing a land with high hills and deep valleys.

Non-marine beds of middle Cretaceous age occur in local areas of subsidence in Westland, Southern Fiordland (Puysegur Point), and coastal Otago. These beds consist of gravels, rock debris, and swamp materials (the last now turned into coal), derived from rugged rapidly eroding landscape. The areas of subsidence are related to the rifting movements then occurring around the margins of New Zealand, but unlike those flooded by the sea, these areas were filled with a wide range of alluvial material.

Middle Cretaceous marine faunas were dominated by bivalves, particularly by the strongly ribbed shell *Inoceramus*. Some specimens of *Inoceramus* reach lengths of 1.5 m and in this respect are like some of the modern giant clams. Other bivalves were like the modern oysters, scallops, and mussels. Ammonites and belemnites were reasonably common; brachiopods and gastropods less common. Cool-water corals, that lived in small clusters on the sea floor, but did not form reefs, are also sometimes preserved as fossils. Large marine reptiles lived in many coastal waters (Fig. 45).

Microscopic plankton (foraminifera and algae) are reasonably plentiful for the first time in New Zealand rocks.

The non-marine beds contain numerous plant remains and plant fossils also occur in many of the marine beds that were deposited close to land. Spores and pollen, stems, leaves, and flowers of plants have been found, derived from ferns, horsetails, clubmosses, cycads, podocarps, and *Araucaria*. The first pollen from advanced flowering plants (Angiosperms) appears in middle Cretaceous rocks at a level corresponding to about 100 million years ago.

The illustrations on the upper margin of the map (Fig. 44) depict (clockwise from top left): the belemnite *Dimitobelus*; the ammonite *Otoscaphites*; the bivalve *Inoceramus concentricus*; southern beech (*Nothofagus*); rewarewa (*Knightia excelsa*). The squid-like belemnite *Dimitobelus*, and the forest trees southern beech and rewarewa, represent the organisms that arrived in New Zealand in mid Cretaceous times via the migration routes that existed around the southern continents. *Otoscaphites* and *Inoceramus concentricus* represent marine organisms that spread from the Tethys Ocean by means of floating larvae.

Fig. 45 Large marine reptiles lived in the seas that covered most areas of New Zealand during Triassic, Jurassic, and Cretaceous times. Fragments of their skeletons have been preserved in the rocks deposited on the sea bed during these times. The remains of large carnivorous lizard-like mosasaurs and long-necked plesiosaurs have been found in rocks of late Cretaceous age, laid down 70–65 million years ago, at a number of localities in northern Canterbury, Marlborough, and a locality in a tributary of the Te Hoe River, inland Hawkes Bay. Reconstructions of two of New Zealand's late Cretaceous marine reptiles are shown in this drawing; the mosasaur *Prognathodon waiparaensis*, about 15 m long; the plesiosaur *Mauisaurus haasti*, with a neck about 6.5 m long and an overall body length of about 14 m.

LATE CRETACEOUS (95–65 million years) Fig. 46

Opening of the Atlantic and Indian Oceans continued at a steady rate throughout late Cretaceous time. The Mozambique Channel opened up between East Africa and Malagasy. The opening movements affecting many sectors of southern Gondwanaland now had the effect of swinging New Zealand northwards from the South Pole, so that in the late Cretaceous it came to lie between 65–55°S latitude (compared with 85°S in the middle Cretaceous; compare Fig. 44).

Although the same movements had also rotated West Antarctica so that it now straddled the South Pole, nonetheless conditions were evidently still not optimal for the accumulation of ice, as traces of late Cretaceous glaciation are unknown. On the contrary, there is abundant evidence from plant beds and fossil wood preserved in West Antarctica that large areas of Antarctica were probably clothed in dense forests of *Nothofagus*. Therefore, as in early and middle Cretaceous times, the climate of New Zealand and other "southern" lands was cool- or cold-temperate and (judging from tree rings preserved in fossil wood), had well-defined seasons.

Rifting movements that continued around the primaeval New Zealand landmass throughout early and middle Cretaceous times culminated about 85 million years ago in the establishment of open ocean in the Tasman Sea and south of New Zealand. Open oceanic conditions brought to an end any likelihood of land routes into New Zealand and New Caledonia from the remainder of Gondwanaland. Shallow-water marine links to the north and west were probably also severed, but New Zealand and New Caledonia were still linked by shallow-water marine routes. Late Cretaceous fossils from New Caledonia are, in most instances, identical with those from New Zealand. They indicate how close the two countries were at this time.

Marine routes extending from New Zealand to southern South America via West Antarctica were also still evident, enabling New Zealand, New Caledonia, West Antarctica, and southern South America to share a "southern" marine fauna (indicated on the map by the diagonally striped pattern). Some elements of the southern marine fauna also extended to southern India, Malagasy, and southern Africa.

Representatives of the "southern" marine faunas that populated the New Zealand region in late Cretaceous times are illustrated on the upper margin of Fig. 46. The illustrations depict (clockwise from top left): the belemnite *Dimitobelus*; the ammonite *Kossmaticeras*; the bivalves *Pacitrigonia* and *Inoceramus pacificus*; the gastropod *Conchothyra*.

The marine links between New Zealand, West Antarctica, and southern South America, although very strong, were not accompanied by the terrestrial links evident in early and middle Cretaceous times. (Those routes used by the ancestral protea groups and by *Nothofagus* to enter New Zealand and New Caledonia.) This is known because in the late Cretaceous ancestral marsupials probably spread from South America to Australia, presumably by using Antarctica as a stepping stone, but apparently were not able to reach New Zealand. The little silhouette drawings placed on the map (Fig. 46) suggest the possible route followed by the ancestral marsupials. Entry into New Zealand was barred by the stretches of open sea then present in the Tasman, Southwest Pacific, and Southern Ocean.

The marsupials (or pouched mammals) include opossums, koalas, wombats, and kangaroos. Today they are found in abundance only in Australia, although opossums are found in North America, and opossums and selvas in South America. Fossil marsupials are known from these countries as well as western Europe, north Africa, and West Antarctica, showing they were formerly more widespread and have been progressively replaced by other more advanced mammals. Although they are otherwise efficient and well-adapted animals, in terms of their reproductive systems monotremes and marsupials may be considered to represent early stages of mammal development. The egg-laying monotremes (platypus and echidna) can thus be viewed as being one step removed from reptiles. The marsupials, on the other hand, represent a further stage of development. Young marsupials are born at a very early stage and crawl into the mother's pouch, where they become attached to a teat and continue growth. This development in the pouch distinguishes marsupials from the more advanced "placental" mammals where development of the young is internal, in a uterus or womb, nourished by a placenta.

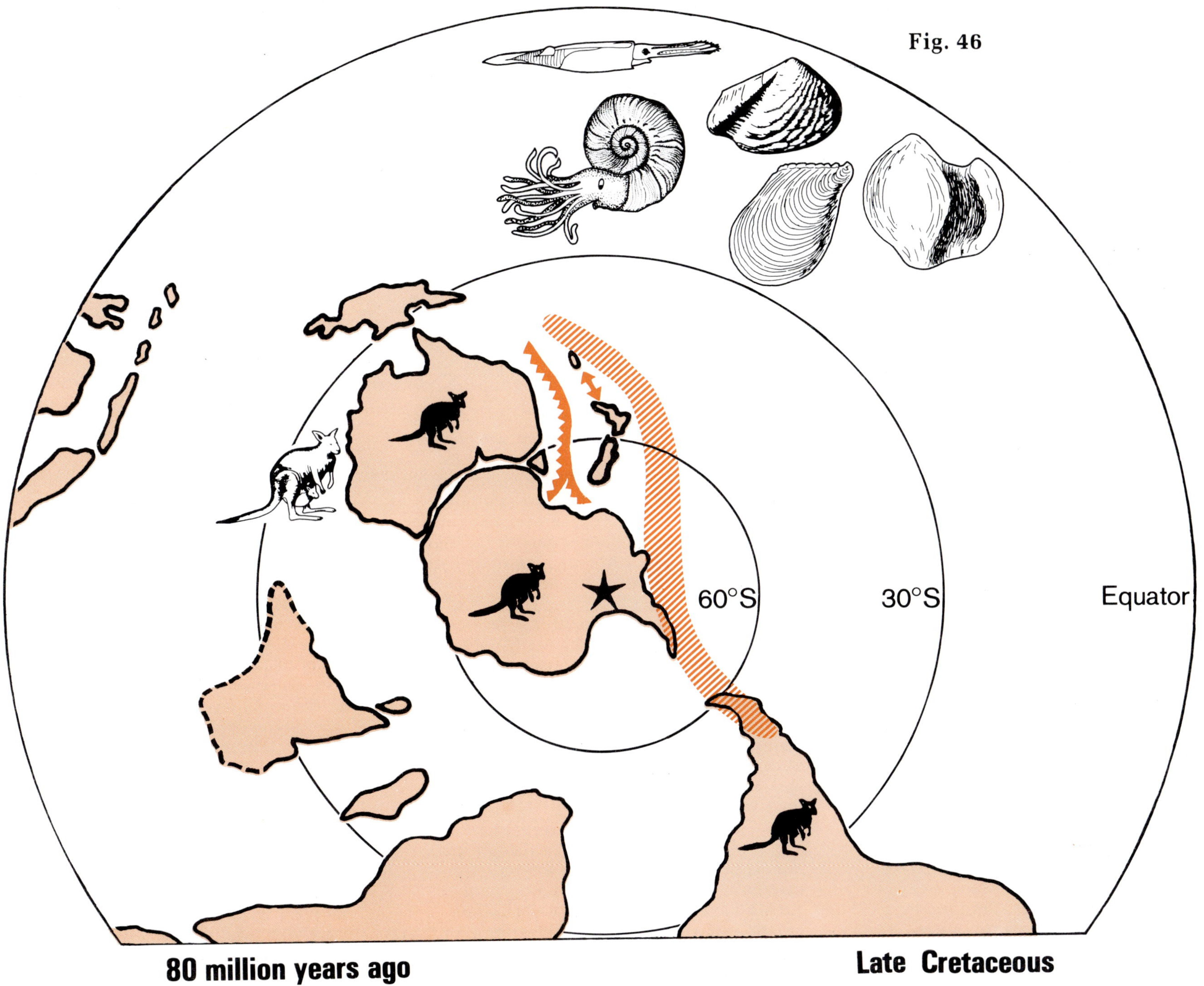

Fig. 46
60°S
30°S
Equator
80 million years ago
Late Cretaceous

THE GREAT DIE-OUT

The end of the Cretaceous, 65 million years ago, saw a dramatic die-out of many of the animal and plant groups that had dominated so many Mesozoic environments. Other periods of extinction had occurred in the course of geological time (for example, at the close of Cambrian, Devonian, Permian, and Triassic times), but the magnitude and apparent suddenness of the Cretaceous extinctions have tended to attract the fascination of the general public and scientists alike. Another factor which focussed attention on Cretaceous extinctions was that dinosaurs, after completely dominating land life for some 175 million years, suddenly died out, never to be seen again.

It is important to remember, however, that the dinosaurs were not the only living things to die out 65 million years ago. In the air the flying reptiles disappeared, but birds and flying insects appear to have been largely unaffected. On the land several groups of mammals became extinct, particularly some of those related to opossums. Other mammals, as well as small reptiles, salamanders and frogs, insects, fresh water clams, and snails seem to have survived unscathed. Some changes also occurred in land plants, but not to the same extent as in the land animals. While some flowering plants became extinct, ferns and cone-bearing trees were largely unaffected.

In the sea many inhabitants of the plankton died out, including many of the types of microscopic calcareous algae (coccoliths) and many of the microscopic single-celled animals called foraminifera. Many extinctions also occurred amongst the larger marine animals. All of the ammonites (cousins of the modern pearly nautilus) and the squid-like belemnites, and all of the giant swimming reptiles disappeared. Thinning-out and extinctions occurred in shellfish (both bivalves and gastropods), brachiopods, bryozoa, corals, sand dollars, starfish and sea eggs, crustacea, sharks, and sea turtles. However, the groups that survived, although sometimes with greatly reduced numbers, later expanded to form the nucleus of modern sea life.

It has been estimated that three-quarters of all living creatures died out at the close of Cretaceous time. Why so many died, while others were little or not affected, is one of the most baffling aspects of the 65 million-year-old mystery.

The fact that creatures died out simultaneously in the air, on land, and in the sea means we must look for a truly world-wide cause or set of causes affecting a wide spectrum of ancient life. Explanations applying solely to the dinosaurs, for example, such as the dinosaurs dying of constipation because of dietary changes forced on them by changes in plant life, or mammals eating dinosaur eggs, simply will not do.

The various global explanations proposed fall into two groups: either earthly changes were responsible, or the causes were beyond the earth and involved cosmic forces. The earthly changes envisaged include changes in climate (heating or cooling), changes in sea level (drastically altering the distribution of land and sea), and volcanic activity (producing a veil of dust in the atmosphere and blocking out the sun). Cosmic forces that have been suggested include the effects of high levels of cosmic radiation, during a time when the magnetic belt normally shielding us from such effects was temporarily weakened; creation of a dust barrier against sunlight; or poisonous rain, perhaps following a collision between a large asteroid and the earth.

A recent theory that has attracted a lot of publicity involves an asteroid some 10–15 km in diameter colliding with the earth 65 million years ago. This theory is based on observations made in various parts of the world (including Marlborough, New Zealand) showing that unusually high (but still minute) concentrations of platinum metals (platinum, iridium, osmium, etc.) occur in thin bands of clay deposited at the very end of Cretaceous time. Such metals, especially iridium, are exceedingly rare on the surface of the earth, but are found in high concentrations in cosmic materials from outer space. It was concluded, therefore, that the high concentrations present in the 65-million-year-old clay could only have come from a source out in space and brought to earth as an asteroid or meteorite.

The theory proposes that the collision of the asteroid formed an enormous crater, perhaps some 150 km across. The dust particles from the crater were thrown high enough into the atmosphere for winds to scatter this material all around the globe, preventing sunlight from reaching the earth. As a consequence, many plants died on both land and sea and many of the animals feeding on them died also. Only small scavenging animals feeding on the dead animal and plant life would have managed to survive until the skies cleared. Among the plants, only those which could regenerate from buried root systems, or those with resistant long-living seeds, would have survived. A variation of this theory suggests instead that a comet passed close to the earth, heating up the atmosphere, and killing many of the large animals and plants. Reaction of cometary material with gases in the atmosphere may have also produced a rain of acids and poisons, such as cyanide, decimating much of the aquatic life.

Fig. 47   New Zealand's only dinosaur is known from a single vertebral bone occurring in marine rocks deposited 70–65 million years ago and outcropping in a tributary of the Te Hoe River, inland Hawkes Bay. The dinosaur was probably about 4 m long, stood about 2 m high and weighed about 0.4 tonnes. The Hawkes Bay dinosaur was a land dweller. However, because the vertebral bone was found in marine sediments along with the remains of ammonites, belemnites, shellfish, mosasaurs, and plesiosaurs (Fig. 45), it probably fossicked along the shoreline, as shown in this reconstruction. If it was a beachcomber, its diet may well have included stranded ammonites because, during late Cretaceous times, they were almost as common as shellfish on today's beaches. Ammonites commonly reached shell diameters of 300 or 400 mm, with a succulent fleshy body protruding to at least an equivalent distance.

Although the "big bang", or "lights out", idea certainly has dramatic appeal, scientists are very wary about invoking "catastrophes" to explain natural phenomena. This is not to deny that an asteroid may have struck the earth 65 million years ago and may have had some effect on the creatures then living. But many other detrimental things were also happening in the natural environment at the time and they probably added to the disaster for many plants and animals. Climate, for example, was changing and undoubtedly there were plants and animals that could not cope with the changes.

The continents of the earth were moving apart at the time and as oceans deepened between them world sea levels dropped. The falling seas changed climatic patterns, oceanic circulation, and salinity, and removed large chunks of living space from the coastal creatures upon whom so many animals depend for food. As the continents moved large splits appeared in the earth's crust and widespread volcanic activity broke out, generating clouds of volcanic ash and dust and causing extensive loss and disruption of living areas.

Thus it is possible that many earthly changes were putting stress on animals and plants, and that many groups, including the dinosaurs, may have been already on a downhill slide towards extinction. Perhaps something like an asteroid striking the earth may have been the "last straw" for many groups of animals and plants. We probably won't ever know for certain. What ever happened, the way was opened for mammals, and eventually humans, to take the centre of the world's stage.

LATE CRETACEOUS IN NEW ZEALAND

By late Cretaceous times erosion of the old New Zealand continent had proceeded to such an extent that long embayments of the sea were fingering in across the land. Such marine incursions were still essentially confined to the edges. For this reason, late Cretaceous rocks, like those of early and middle Cretaceous, are predominantly found in areas around the edges of the modern landmass (Fig. 48). Some of the marine embayments were simply extensions of areas already flooded by the sea in middle Cretaceous times. In these instances late Cretaceous rocks are laid down directly on top of middle Cretaceous rocks, indicating continuous marine conditions. These areas include the Raukumara Peninsula, Wairarapa, northern Marlborough, and some parts of Northland.

Other areas of New Zealand were newly invaded by the sea in late Cretaceous times and marine sediments of this age are directly laid down across the eroded remains of older rocks. Sequences in southern Marlborough, northern Canterbury, southern Hawke's Bay and some parts of Northland were deposited in this way. Mudstone is the most abundant rock type, with lesser amounts of sandstone and limestone.

The late Cretaceous marine faunas of New Zealand, while generally comparable to those of the middle Cretaceous, are more diverse, better preserved, and often richer numerically. The faunas are fully representative of the marine life of the time. As in the earlier part of the Cretaceous, large coarsely ribbed clams (*Inoceramus*) remain important members of the sea-floor community.

Compared with the early and middle Cretaceous, ammonites become more abundant in the late Cretaceous, and belemnites in some environments. A great variety of bivalves and gastropods is also commonly preserved. Other groups represented in late Cretaceous fossil assemblages are: nautiloids, sponges, corals, bryozoa, brachiopods, marine worms, scaphopods (tusk shells), crustacea, barnacles, starfish, sea eggs, and sea lilies. Turtle remains are known from Hawke's Bay, and sharks teeth are common in many localities. (As the skeleton of a shark is composed of cartilage it is not normally preserved, and only the teeth remain as fossils.) Skeletons of bony fish and isolated fish scales are also known.

Major discoveries of large marine reptiles have been made in Hawke's Bay and north Canterbury. These include long-necked plesiosaurs and the fearsome giant aquatic lizards called mosasaurs (Fig 45).

Hawke's Bay is also the site of the only known occurrence of a dinosaur in New Zealand. The single vertebra, from the latest Cretaceous, 85–70 million years ago, is from a terrestrial carnivorous dinosaur, about 4 m long and weighing about 0.4 tonnes. Such dinosaurs, known from all continents, except Antarctica, undoubtedly came to New Zealand at about the same time as the ancestors of the moa and kiwi (see Fig. 40) and furthermore, may well have been one of their main predators. Other land prey may have included the tuatara. Food may also have been obtained by means of fossicking along the sea shores, feeding on dead and dying marine creatures, especially the ammonites that were probably reasonably plentiful at the time, and undoubtedly provided a succulent meal! (See Fig. 47.)

The plankton of the late Cretaceous seas included many types of foraminifera, and calcareous algae (coccoliths).

The close proximity of land in the late Cretaceous is indicated by the frequent occurrence of fossil plant material (wood, leaves, etc.) mixed into the marine sediments, and by the presence of coal beds of this age in Northwest Nelson, the West Coast, and South Otago.

GREAT CRETACEOUS DIE-OUT IN NEW ZEALAND

Although sea continued to cover large areas of
New Zealand during the time of transition from
Mesozoic to Cenozoic, 65 million years ago, the
continuity of the fossil record is often broken.
Local earth movements resulted in beds of beach
gravel or river-laid material, or actual erosion
gaps. However, continuously fossiliferous
sequences spanning the Mesozoic/Cenozoic
boundary are available in some areas, notably Te
Uri Stream (Dannevirke area), the Ward area and
Clarence Valley (Marlborough), Waipara River
and Haumuri Bluff (north Canterbury) and the
Dunedin–Shag Point area (Otago). In these areas
complete successions of marine fossils can be
traced from the late Cretaceous across the
boundary and beyond into the Paleocene.
Microfossils occur in the sediments of all the
areas and successions of tiny calcareous single-
celled animals (foraminifera), and plants
(coccoliths and dinoflagellates) are available for
study. Many members of these groups show
sudden extinctions at the boundary.

Large fossils occur in abundance in the Clarence
valley, Waipara River, and Haumuri Bluff. In the
middle course of the Waipara River and adjacent
tributary streams, fossils of bivalves, gastropods,
and squid-like belemnites can be traced up to
the boundary. Bones of large marine reptiles
(mosasaurs and plesiosaurs), encased in
concretions (large rounded mineral accumulations
looking like enormous cannonballs), are also
found extending upwards to very close to the
boundary. All these fossils disappear at the
boundary, giving a New Zealand version of the
Great Die-Out.

Although other, more detailed, measurements are
progressing, preliminary chemical measurements
across the boundary in northeastern
Marlborough, have established that the New
Zealand area was also affected by the event
causing the unusual concentrations of platinum,
iridium and osmium etc.

Fig. 48 Erosion of the ancestral landmass continued at
an inexorable rate during middle and late Cretaceous
time. Towards the close of the Cretaceous, 70–65
million years ago, much of the land was at a low level
and sea was steadily encroaching over the levelled-off
land along the east coasts of both islands and in
Northland. Large swamps that formed in low-lying
areas of the eroded landscape laid down thick peaty
deposits. These produced the coalfields of eastern Otago
(e.g., Kaitangata), some of those of the West Coast
(although most are of mid-Eocene age), and those of the
west Wanganui–Aorere–Collingwood region (northwest
Nelson).

  Land is indicated by the coloured area, the Alpine
Fault by a broken line, volcanoes by cone symbols, and
coal swamps by a swamp symbol. The outline of
modern New Zealand is shown for reference, but it
must be remembered that at this time it did not
physically exist.

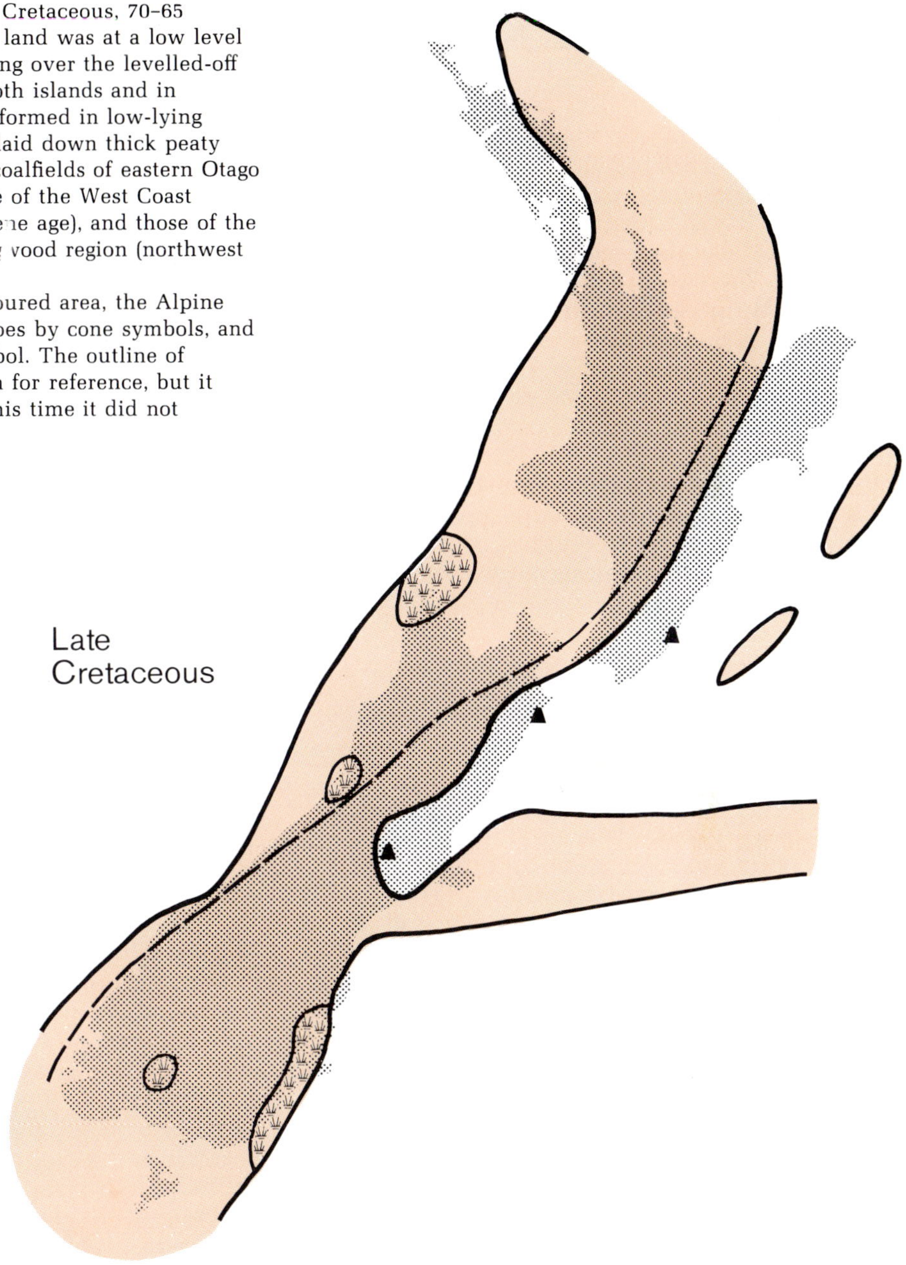

PALEOCENE (65–53 million years ago)   Fig. 49

The global movements that were gradually swinging New Zealand northwards away from the South Pole continued during the Paleocene. These movements, beginning in the late Cretaceous, resulted from creation of new sea floor in the actively opening Indian, Atlantic, and South Pacific oceans. Under this influence New Zealand moved steadily northwards, so that in the Paleocene it occupied a position of 50–45°S latitude, compared with 65–66°S latitude in the late Cretaceous (Fig. 46). The northwards movement of New Zealand into regions influenced by warm-temperate oceanic currents brought about slight warming so that the cool-temperate climates of the Cretaceous gave way to warm-temperate climates, comparable to today. The northerly drift of New Zealand, and continued creation of new sea floor to the south of New Zealand, eventually broke the southern shallow-water marine links between New Zealand, western Antarctica, and South America. Sea floor spreading, starting in the Tasman Sea in late Cretaceous times 80 million years ago, continued into the early Paleocene, but ceased 60 million years ago when the Tasman had probably reached its present width (1850 km). However, some time in the Paleocene, New Zealand probably received the ancestors of the primitive bats that are its only native mammals.

The New Zealand bat fauna comprises two species: the short-tailed bat, *Mystacina tuberculata* and the long-tailed bat, *Chalinolobus tuberculatus*. *Mystacina*, representative of a very ancient bat lineage, probably arrived before distances became too great across the widening oceanic gaps around New Zealand. *Chalinolobus*, on the other hand, is a more advanced type of bat. It is a relatively recen immigrant from Australia, presumably blown across the Tasman by westerly storms, generated by the West Wind Drift that came into being in post-Miocene times.

Like the ancestors of the short-tailed bat, ancestors of three distinctive families of New Zealand land birds may have also arrived some time in the Paleocene as wind-blown immigrants across the steadily widening Tasman Sea and Southern Ocean. These families are found nowhere else in the world, and are presumably long-standing residents. They include the New Zealand wrens (rifleman, bush wren, rock wren, Stephens Island wren); the native thrushes (piopio); and the wattlebirds (huia, saddleback, and kokako). It is perhaps not surprising that the only land-based animals that were apparently able to cross the infant Tasman Sea as it was opening, were representatives of newly evolved groups with powers of flight: bats and birds.

The map (Fig. 49) illustrates the geography of the Southern Hemisphere at a time in the Paleocene, some 60 million years ago, when the Tasman Sea was opening out to its full width and a steadily widening oceanic gap was opening up between New Zealand and Antarctica. The large arrows indicate possible routes taken by newly evolved animals with powers of flight. By island-hopping south from Indonesia they were able to reach New Guinea and Australia, and then were able to cross the infant Tasman Sea before it reached its full width. Birds ancestral to the modern New Zealand wrens, thrushes, and wattlebirds may have come to New Zealand via such a route. While ancestors of the native bat *Mystacina* may also have used such a route, it is also possible that they may have come from South America via eastern Antarctica and Australia.

The drawings comprise (clockwise from the top): the gastropod *Struthiolaria*, representing the marine animals that arrived in New Zealand in Paleocene times via southern routes; the nautiloid *Aturia*, representing the marine animals that probably arrived via northern routes; the wattlebirds kokako (*Callaeas cinerea*) and huia (*Heteralocha acutirostris*), and the New Zealand short-tailed bat (*Mystacina tuberculata*), representing the land creatures that arrived in New Zealand at about this time by flying across the newly opened oceanic gaps.

Sea-floor spreading between New Zealand/Antarctica and Antarctica/South America progressively weakened the shallow water marine links that had existed between these lands in Cretaceous times. However, New Zealand and New Caledonia still continued to share marine faunas, as they had done since at least the Permian.

As creation of new sea floor ended in the Tasman, the focus of sea-floor spreading shifted to the southwest, to between Australia and Antarctica. As a result, a wedge of new ocean floor began to appear between these two continents some 55

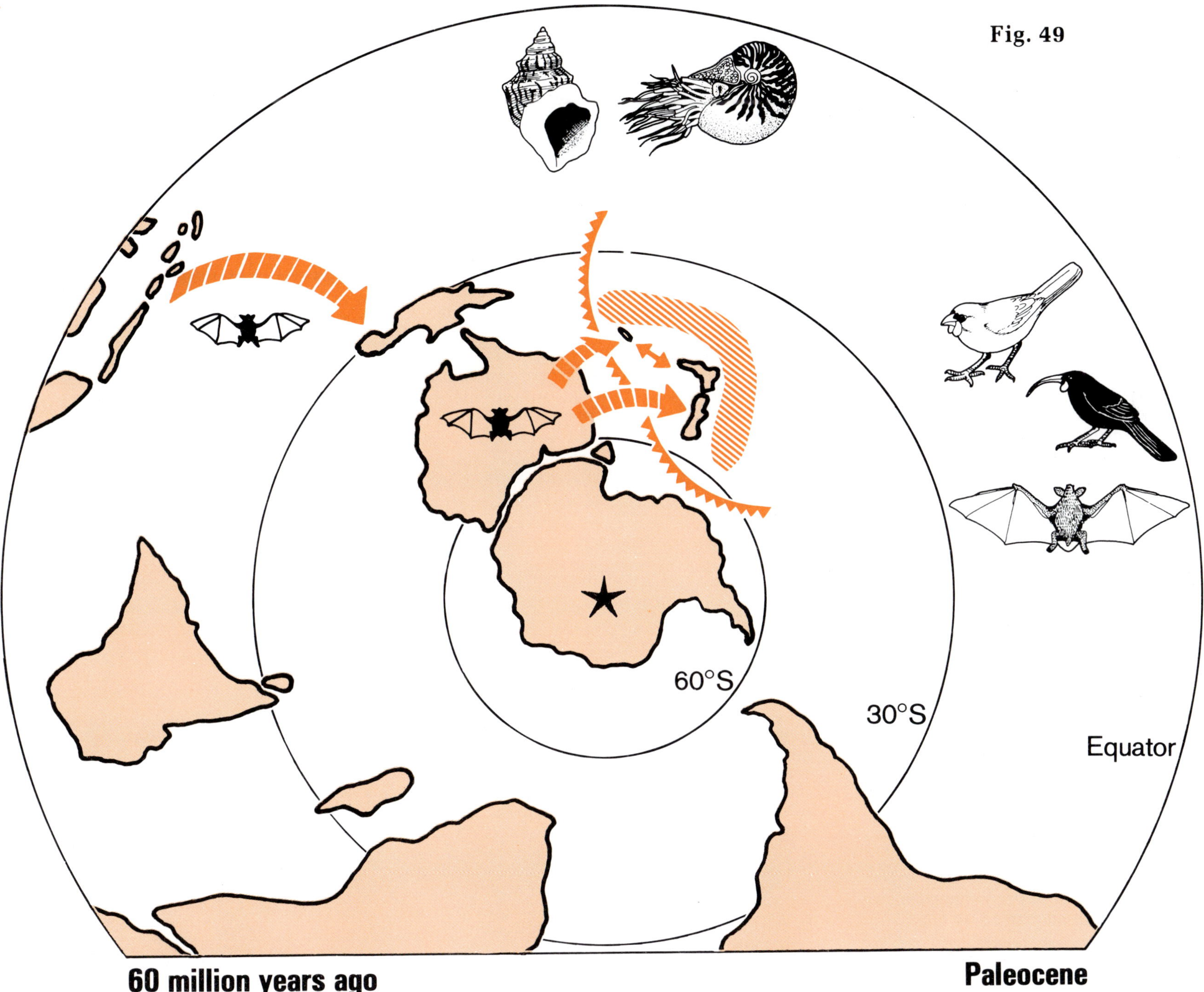
Fig. 49
60°S
30°S
Equator
60 million years ago
Paleocene

million years ago, at the close of the Paleocene. As Australia and Antarctica moved apart, the zone of actively spreading sea floor between them linked up with that already in existence south of New Zealand. Australia and New Zealand thus began to move northwards together, while Antarctica began its journey southwards towards its modern position straddling the South Pole.

PALEOCENE LIFE

The Paleocene is the earliest time division (period) of the Cenozoic era of "recent life" (*cenos* = recent; *zoe* = life). The Cenozoic era is often called "The Age of Mammals". Following the demise of the dinosaurs in the Cretaceous, the mammals came into their own during the 65 million years of this era.

As shown by the origin of its name ("recent life"), the Cenozoic era is also characterised by the progressive evolution of modern forms of life. The gradual increase in the proportion of modern types which occur as fossils in the Cenozoic era is reflected in the names of the various time divisions or periods, all constructed with an appropriate prefix added to -cene (derived from *cenos* = recent). The oldest period, the Paleocene, is derived from *palaios* = old. The next oldest, Eocene, is derived from *eos* = dawn; Oligocene is from *oligos* = few; Miocene from *meion* = less (that is, less than in the next period, the Pliocene). The three most recent periods are called Pliocene (*pleion* = more); Pleistocene (*pleiston* = most); Holocene (*holos* = complete, whole). The comparative terms (less, more, most, complete) refer to the relative proportions of modern marine shells occurring as fossils.

The massive waves of extinctions that occurred at the close of the Cretaceous left a great number of vacant ecological niches in a wide range of environments. These niches were quickly taken over by the surviving groups of animals and plants. Mammals took the place of dinosaurs on the land. Ammonites and belemnites were succeeded by squids, cuttlefish, octopus, and relatives of the modern pearly nautilus. Gastropods and bivalves took over many of the ecological niches formerly occupied by brachiopods. Amongst microscopic marine life a rapid expansion of calcareous plankton occurred, following their virtual extinction at the close of the Cretaceous. Many early Paleocene faunas include stragglers from Mesozoic times, survivors of the late Cretaceous extinctions, but these were progressively thinned out as Paleocene time passed and newcomers appeared.

During the "Age of the Dinosaurs", the early mammals probably survived by filling two nocturnal feeding niches, then unoccupied. One niche was that of insect eating (like modern ant-eaters) and the other was that now occupied by rodents. Despite the dramatic changes that have occurred in mammalian groups, the rodent-like way of life has remained unchanged to the present day, making it the longest mammalian lineage, spanning 100 million years.

When the dinosaurs became extinct many of the daytime feeding niches became available and the mammals were able to rapidly take advantage of the situation. On every continent (except New Zealand, isolated before the evolution of most mammal groups), certain basic types of mammal evolved, filling the major ecological niches of rodents, insectivores, herbivores, and carnivores. Some of the early mammal faunas became isolated on separate continents for long periods of time. Until the arrival of the Aborigines and the dingo, 40 000 years ago, marsupials and monotremes made up Australia's mammal population. Unusual mammal faunas also evolved in isolation on Africa and South America, until land connections were re-established in late Miocene (Africa) and late Pliocene (South America).

PALEOCENE IN NEW ZEALAND

Erosion of the old landmass dating back to late Jurassic and earliest Cretaceous continued throughout the Paleocene. Rivers and streams wore down the hills and the sea ate into the coastline (Fig. 50). Areas that had already been submerged in Cretaceous times became more deeply and more extensively submerged in Paleocene times. These areas included Northland, the eastern flanks of the North Island, coastal Marlborough, and northern Canterbury. Other newly submerged areas included coastal southern Canterbury and Otago. Sea also lapped over the areas now occupied by the modern Chatham Islands, Campbell Island, Auckland Islands, and the Lord Howe Rise.

Marine fossils of Paleocene age occurring in New Zealand include descendants of Cretaceous groups of bivalves and gastropods, survivors of the end-Cretaceous extinctions. Newcomers include bivalves, gastropods, and sea eggs, derived from northern sources. A fossil turtle of Paleocene age is known from Ward, Marlborough. Plant fossils include ancestors of the southern beech (*Nothofagus*), pohutukawa, and she-oak (*Casuarina*). The protea group expanded and diversified.

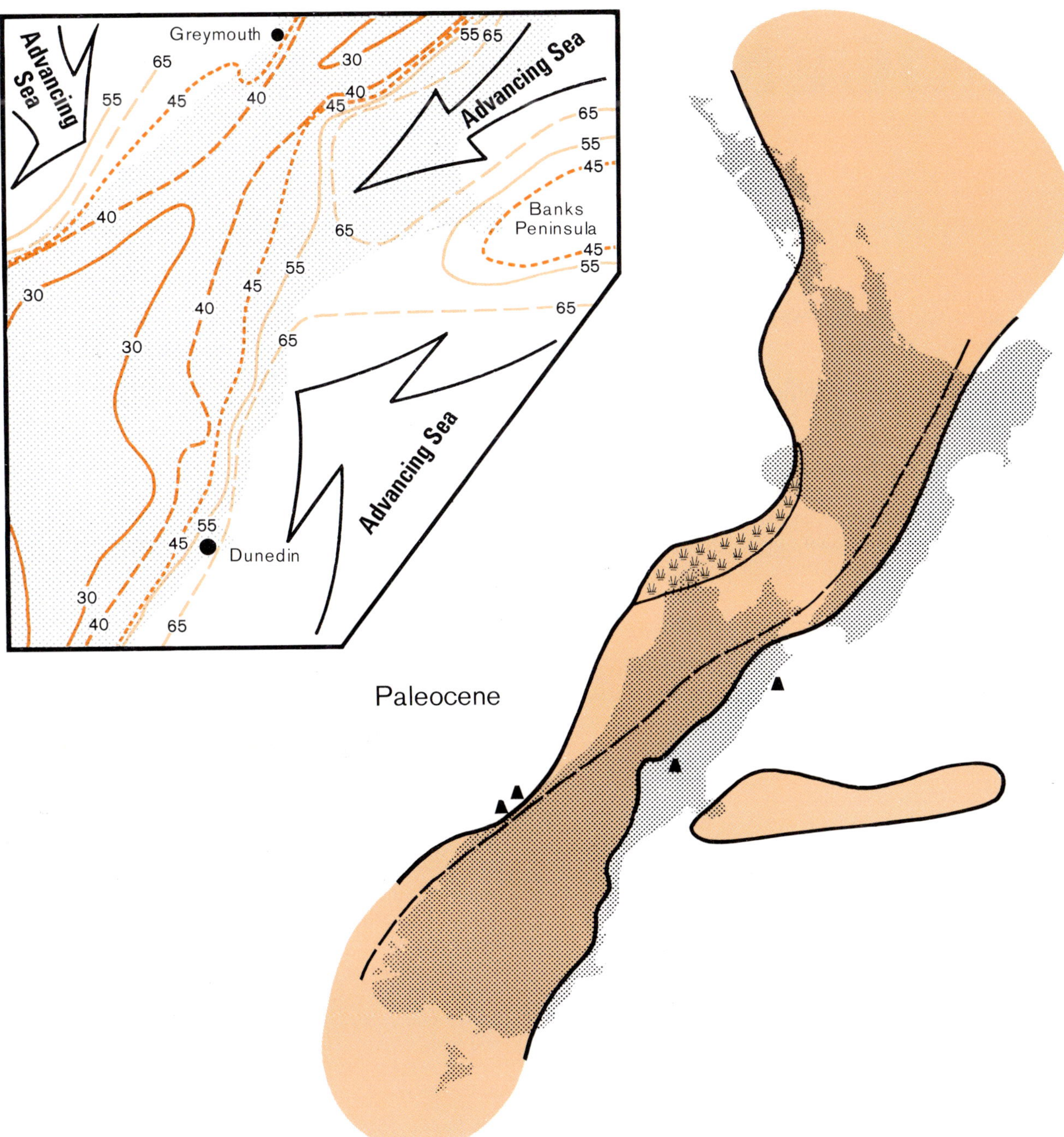

Fig. 50 During the Paleocene (65–53 million years ago), the old New Zealand landmass became lowered still further by erosion. Little if any high country remained and extensive plainlands were developed. The sea continued to inundate low-lying areas around the margins of the landmass, extending further westwards on the east coasts of both islands, and started to encroach on the west coasts. Large swamps formed on low-lying parts of the landscape in what is now the western approaches to Cook Strait. Coals formed from these swamps and later ones (mid Eocene) have been found in oil-prospecting drill holes off-shore from Taranaki and Nelson. It has been suggested that the oil and gas of the Kapuni and Maui fields originated as a by-product of the coal-forming processes.

The northern half of the South Island provides an excellent example of the gradual encroachment of the sea over the land that occurred before, during, and after Paleocene times. Detailed geological mapping has identified a series of shorelines traversing the region in a roughly NE–SW direction, corresponding to stages in the advance of the sea across the worn-down land during Cretaceous and early Cenozoic times. The positions of these shorelines at 65, 55, 45, 40, and 30 million years ago are shown on the accompanying inset map (top left).

Land is indicated by the coloured area, the Alpine Fault by a broken line, volcanoes by cone symbols, and coal swamps by a swamp symbol. The outlines of modern New Zealand are shown for reference, but it must be remembered that at this time it did not physically exist.

EOCENE (53–37 million years ago) Fig. 51

Creation of sea floor in the Southern Ocean south of Australia and New Zealand continued to move Antarctica southwards and Australia and New Zealand northwards. As a consequence, New Zealand's geographic position in the late Eocene was 45–40°S latitude, compared with 50–45°S latitude in the Paleocene (compare Fig. 49).

As Antarctica moved into higher latitudes it gradually lost its role as a stepping stone for southern migrants. In Paleocene, Eocene, and Oligocene times (65–24 million years ago), many parts of Antarctica were covered by beech forests similar to those in New Zealand and southern South America today. For example, throughout this period coastal areas of West Antarctica probably provided potential migratory routes for cold-temperate organisms. However, as Antarctica moved southwards, and cold marine currents began to flow around the continent, ice fields formed on the mountain tops and glaciers began to reach down the valleys towards the sea.

Initially, the Antarctic beech forests could probably live alongside the glaciers and ice fields, as they do today in Patagonia and southern New Zealand, but eventually further chilling resulted in massive expansion of ice, and retreat of the plant life. With nowhere to retreat to, because links to New Zealand and South America were broken, forest life eventually became extinct on Antarctica. Certainly all indications are that by Pliocene times at the latest (5 million years ago) Antarctica had virtually been submerged under a thick ice cap, signalling the end of forest life on the frozen continent. Judging from deep sea drilling offshore from Antarctica, ice first started to lie on Antarctic mountain tops in the early Eocene (53–50 million years ago). Then in the latest Eocene (40–37 million years ago), a major cool phase occurred and tongues of ice reached the Antarctic coast. The first sea ice formed and the waters around Antarctica became cold. It was not until the middle Miocene, however, (16–13 million years ago), that the first major ice sheets became established in Antarctica.

As Antarctica moved closer to the South Pole, seaways opened up around its coastline. The first penguins appeared in the late Eocene in southern Australia, southern New Zealand, and Seymour Island (Antarctic Peninsula), but they appeared later in other southern lands. These early penguins, while including some ranging up to the size of the modern emperor penguin (1 m in height and weighing 30 kg), also included a giant New Zealand species *Pachydyptes ponderosus*, recorded from late Eocene strata at Oamaru. *P. ponderosus* had a height of 1.62 m and weighed a massive 100 kg.

As New Zealand and Australia moved slowly northwards because of steady expansion of sea floor in the Southern Ocean, oceanic temperatures gradually began to rise. In the Paleocene, sea temperatures around ancestral New Zealand were roughly equivalent to those now found around modern New Zealand, while in the Eocene they became warmer, even subtropical, for a time. However, as ice began to build up on Antarctica, a decline in sea temperatures became apparent around New Zealand in the latest Eocene and earliest Oligocene.

The warm Eocene seas lapping over much of ancestral New Zealand encouraged many tropical and subtropical marine animals to move south from Southeast Asia, Indonesia, and eastern Australia. Because there were no shallow-water links, the organisms that migrated south into New Zealand waters at this time and later, were types capable of being widely distributed across open ocean. That is because the adults, eggs, or larvae could float or swim.

The opening of the Southern Ocean also progressively weakened the shallow-water marine links between New Zealand, Antarctica, and South America. Such links had become rather tenuous by late Eocene time and disappeared altogether after the end of the Eocene. The "southern" mollusca that were shared by New Zealand and South America at this time were of types that probably had a free-swimming larval stage. Examples are *Speightia* and *Struthioptera*, illustrated (top right) on Fig. 51. Their distribution was probably assisted by the oceanic current systems then starting to develop around Antarctica.

At about the same time in the late Eocene, opening of oceanic basins to the north of New Zealand broke the long-standing continuity between New Zealand and New Caledonia. From almost the beginning of their geological histories, New Zealand and New Caledonia shared geological events and may have originally been part of a single

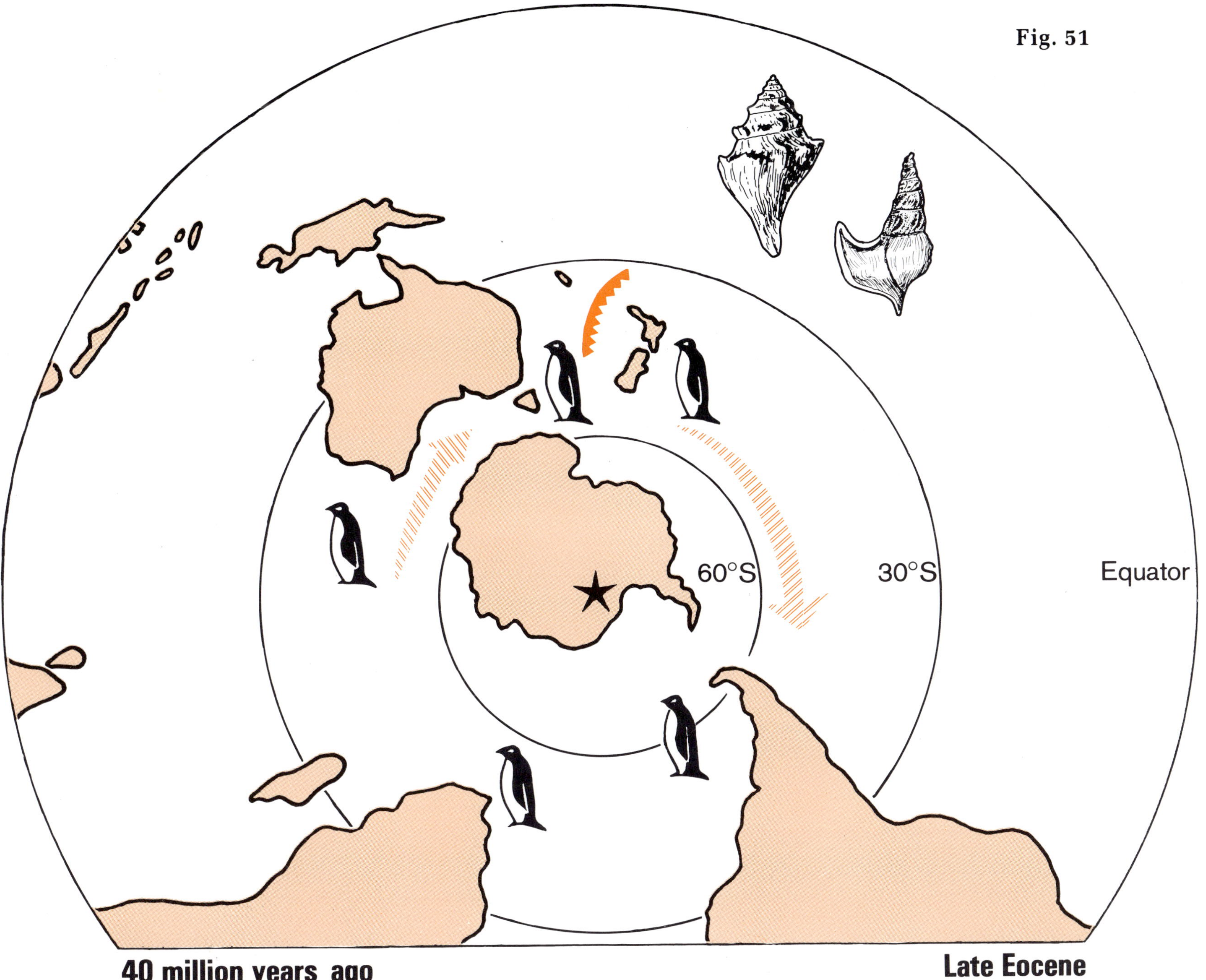

Fig. 51
60°S
30°S
Equator
40 million years ago
Late Eocene

ancestral landmass. New Caledonia probably also shared many elements of New Zealand's primitive terrestrial flora and fauna and although "thinning-out" has occurred over geological time, geographic isolation has ensured that many interesting relics remain. However, towards the end of the Eocene, sea floor movements progressively rifted New Zealand and New Caledonia apart, and from then onwards both countries went their separate ways.

All available evidence suggests that from late Eocene time onwards New Zealand was completely surrounded by deep ocean. There are no signs of land or shallow water links to neighbouring countries as had been present in Jurassic and Cretaceous times.

EOCENE LIFE

The Eocene (and Paleocene) were times of great richness in mammal development. During this time many forerunners of modern animals came into being, but they were markedly unlike the species of today. Ancestral forms of the horse, for example, were no bigger than a fox terrier. Forebears of the modern rhinoceros were about the size of a small horse. Also among the mammal groups to diversify in Paleocene and Eocene times were forms ancestral to the modern primates. Like all of the early mammals, the first primates were probably nocturnal insect eaters in the Cretaceous, but in the Paleocene and Eocene they occupied the niche of tree-living gnawers and nibblers, rather like squirrels.

In the marine realm, most of the Cretaceous stragglers had disappeared before the close of the Paleocene. The Eocene was characterised by great expansion and diversification of modern groups. Planktonic organisms flourished in many parts of the world and large foraminifera were abundant in tropical seas. The niches left vacant at the end of the Cretaceous by the extinction of large marine reptiles (such as mosasaurs, ichthyosaurs, and plesiosaurs) were filled by ocean-going mammals. One of the first groups to appear was the cetaceans (whales, dolphins, porpoises). The earliest ancestral cetacean fossils occur in the early Eocene of Pakistan. Their characteristics suggest they were ungulate (hoofed) animals which were making a gradual change from land into the sea, spending more and more time feeding on plankton and herring-like fishes in shallow waters of the eastern Tethys Ocean. These ancestral cetaceans were not fully adapted for an entirely marine existence, and lacked modifications for directional hearing under water, for example.

EOCENE IN NEW ZEALAND

Erosion of the old New Zealand landmass continued at an inexorable pace throughout the Eocene. By the end of the period many areas that had previously been land of diverse relief were beginning to be worn down to virtually featureless plains, lying close to sea level. In some areas the wearing-down process had reached such an advanced stage and the land was at such a low level that sea flooded in (Fig. 50). Elsewhere, extensive swamps formed in low-lying areas of the worn-down landscape, notably in Northland, Waikato, Southland, and along the West Coast of the South Island. The plant material laid down in these swamps became the basis of the coal fields now found in these areas. Most of these coal deposits were flooded by the sea in the middle Eocene. In the Greymouth area, for example, sea transgressed over the entire area and extended northwards to Nelson, but left a large island in the Karamea region in Northwest Nelson. Sea was not far offshore from the present Taranaki coastline. On the east, deep water limestones formed in Marlborough, North Canterbury, Campbell Island, and on the Chatham Rise.

For the first time during the gradual wearing down and submergence of the old landmass, the sea began to invade the land on all sides, towards the end of the Eocene (Fig. 52). New incursions of the sea occurred in Northland and Southland and along the east and west coasts of both islands. Extensive coal swamps formed in Waikato and in Taranaki. Sea extended further inland in Marlborough and Otago, almost meeting the sea coming in from the West Coast. A large marine trough developed on the edge of Fiordland, in the Waiau–Manapouri–Te Anau region, although Fiordland itself and the remainder of Otago and Southland remained as land (Fig. 52).

New Zealand Eocene marine fossils include the first members of many distinctive bivalves, gastropods, and echinoderms that continued to dominate the marine faunas for a large part of Cenozoic time. A notable Australian element is evident in New Zealand Eocene marine faunas, indicating the presence of marine currents originating from Australian sources. Warm-temperate conditions are indicated by early Eocene faunas, but the presence of large foraminifera and corals at times in the middle and late Eocene indicate sub-tropical conditions. Few of these warmth-loving newcomers survived the sharp drop in temperature at the Eocene/Oligocene boundary.

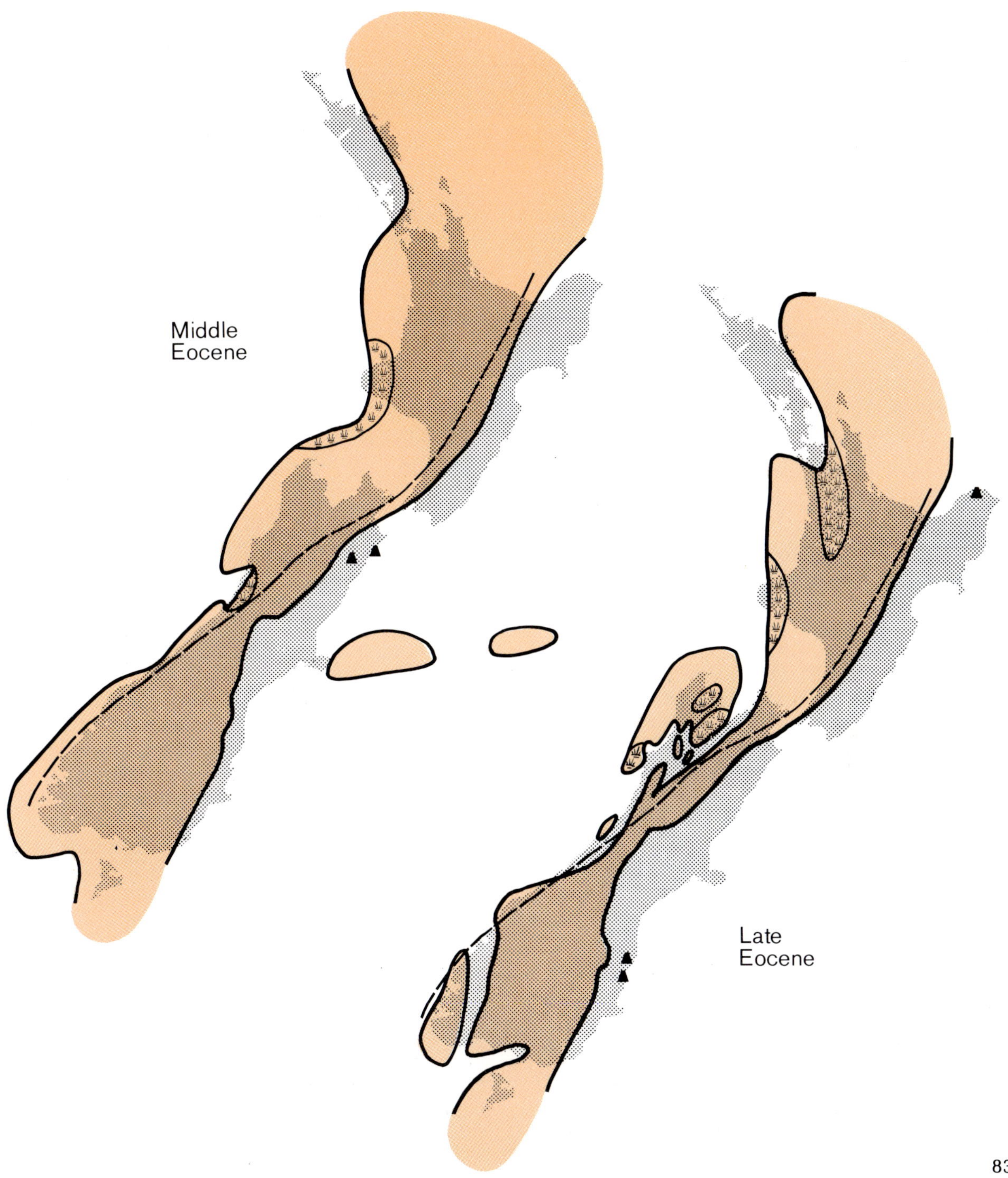

Fig. 52 By the middle Eocene (45 million years ago), as depicted in the left-hand diagram, erosion of the old landmass had proceeded to such an extent that deeply eroded plainlands were widespread throughout New Zealand. As the land became progressively lower, the sea encroached further on the east and west coasts of both islands. The main coalfields of the Westport–Greymouth region accumulated at this time in brackish swamps occupying low-lying areas close to the sea.

By the end of the Eocene (about 40 million years ago), most of the old New Zealand landmass had been worn down to a vast plain and was lying virtually at sea level (right-hand diagram). Sea had invaded many areas. In the South Island, sea flooded along a broad valley on the eastern flank of Fiordland, extending from the Waiau Valley to the Hollyford. In Canterbury and Marlborough the sea reached far inland, and in mid Canterbury it almost met the sea extending in from the West Coast. Swamp deposits accumulated in Waikato and Taranaki, forming the basis of coal deposits now mined in these areas. Volcanic islands were actively erupting at East Cape and the Dunedin–Oamaru region.

In both diagrams, land is indicated by the coloured area, the Alpine Fault by a broken line, volcanoes by cone symbols, and coal swamps by a swamp symbol. The outlines of modern New Zealand are shown for reference, but it must be remembered that at this time it did not physically exist.

OLIGOCENE

(37–24 million years ago) Fig. 53

The Oligocene was a time of significant geographic, climatic, and biological changes that set the world on a course leading towards development of many of the environments we know today. Continuing sea-floor spreading movements and creation of new ocean floor in the Southern Ocean immediately south of New Zealand and Australia steadily moved both countries northwards towards mid latitudes (compare Fig. 51). Corresponding movements moved Antarctica progressively southwards.

Widening of the oceanic gap south of Australia and New Zealand, and movement of Antarctica closer to the South Pole, triggered a series of inter-related events, which extended throughout Oligocene time and together helped shape the developing character of the Southern Ocean. These events included progressive accumulation of ice on Antarctica, creation of circum-Antarctic oceanic currents, spread of cool Antarctic water, and influence on southern marine life of mineral nutrients derived from Antarctica by glacial activity.

Southwards movement of Antarctica led to climatic conditions favouring precipitation and accumulation of snow and ice— at first on the high country of inland Antarctica in early and middle Eocene, but later extending outwards as valley glaciers to reach the Antarctic coastline in latest Eocene and earliest Oligocene.

Expansion of ice on Antarctica had a major cooling effect on waters of the Southern Ocean. In latest Eocene and earliest Oligocene, sea water temperatures around

New Zealand dropped from those of the subtropics to those now found around the coasts of the sub-Antarctic islands south of New Zealand (e.g., Campbell and Auckland Islands). These cool-temperate conditions favoured development in the Southern Ocean of various groups of plants and animals that preferred cool habitats.

Build-up of ice on Antarctica was also accompanied by changes in patterns of oceanic currents within the surrounding seas. These changes enhanced the cooling already occurring on Antarctica, and transported cool Antarctic water into many southern regions. Such changes could only take place after the southern lands had moved away from Antarctica. This left southern oceans clear of barriers and allowed free circulation of currents.

Although open sea first appeared around Antarctica in the late Cretaceous (80 million years ago) when Africa and New Zealand began to pull away, followed by South America (70–65 million years ago), it was not until late Paleocene times (55 million years ago) that Australia and Antarctica began to separate— the last remaining southern landmasses to drift apart. At first, as Australia and Antarctica pulled apart, the gap between the two countries was filled by a shallow sea stirred by weak oceanic currents. Such a shallow sea with limited circulation continued to exist to almost the end of the Eocene (between 40 and 37 million years ago), when enough of a gap had been opened to allow more open-sea conditions to develop (Fig. 51). Initiation of the circum-Antarctic current was not

possible, however, until Tasmania, and more critically, its submarine continuation as the South Tasman Rise (the cross-hatched pattern on Fig. 53.), had moved clear of the northern tip of Antarctica's Victoria Land. By latest Eocene time, although some marine life was able to move from the southeast Indian Ocean, across the South Tasman Rise, and into the southeast Pacific Ocean, it is evident that only a very shallow marine connection existed and full separation had not yet been established.

In the early Oligocene the South Tasman Rise still had not moved far enough north-wards, away from the Antarctic coast, to allow free ocean current movement, although open-sea conditions were present on either side of the rise. However, by middle and late Oligocene time (about 34–24 million years ago), a sufficient gap had opened to create oceanic and meteorological conditions leading towards a single unified circum-Antarctic current system (indicated by arrows on the map). Once established, development of the current system was greatly enhanced by the widening gaps between Antarctica, South America, and Australia, allowing increasing quantities of water to be involved in the current movements.

The stirring of the Southern Ocean by the infant circum-Antarctic current system had the effect of spreading cool water, derived from the Antarctic glaciers, northwards into surrounding regions. Thus many southern areas began to feel the effects of influxes of cool Antarctic water during late Eocene and Oligocene times. In New Zealand, for

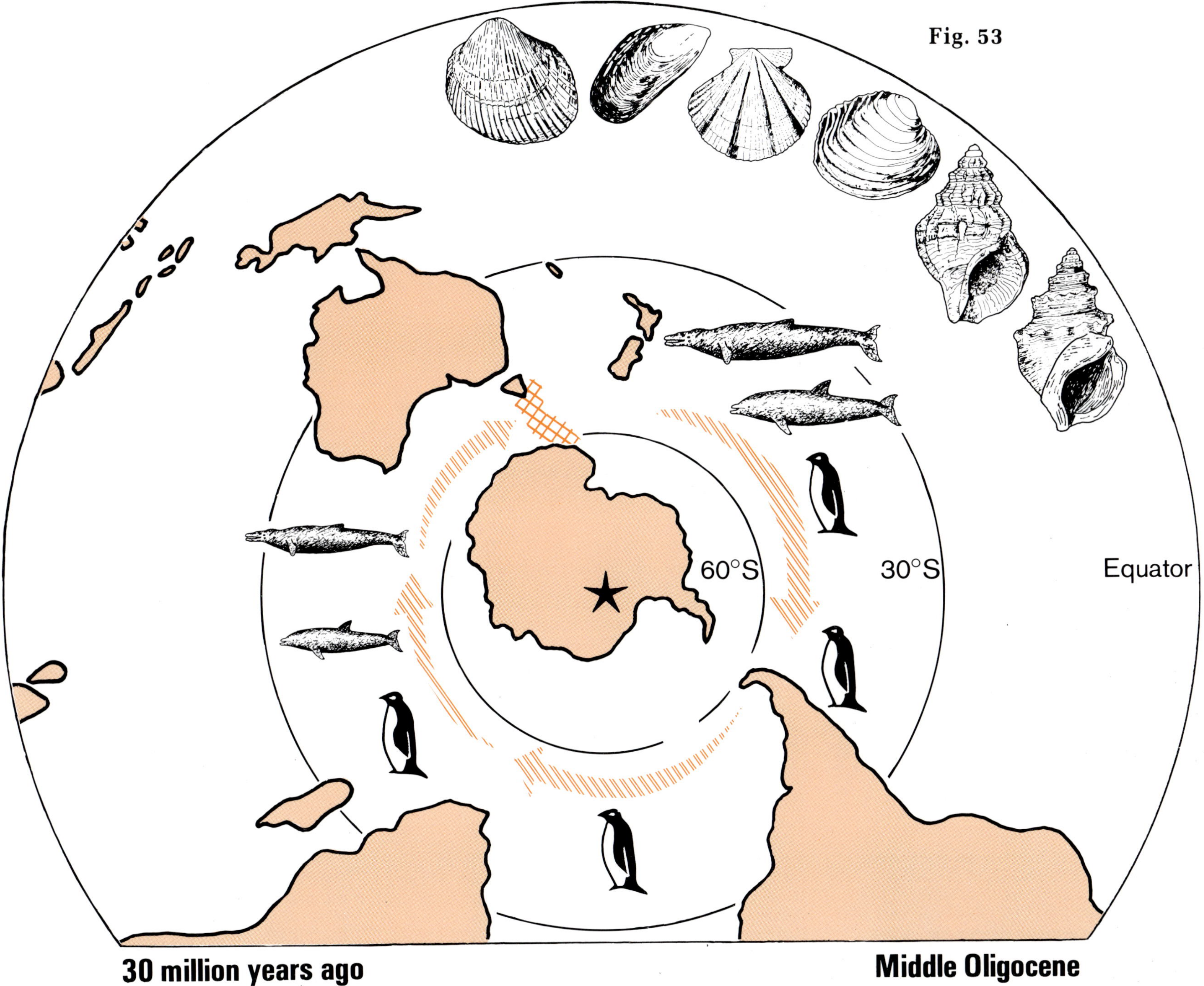
Fig. 53
60°S
30°S
Equator
30 million years ago
Middle Oligocene

example, subtropical environments were gradually replaced by those of a more temperate nature in late Eocene and especially Oligocene times.

As the Antarctic glaciers became more extensive, a variety of rock materials were exposed to the effects of glacial erosion. Scouring, shattering, and general disintegration of rock produced fine mineral powder called "rock flour". The mineral products derived from glacial activity eventually found their way into the surrounding oceans, either from glaciers directly entering the sea, or from rivers draining ice-covered areas inland.

The microscopic plants and animals that make up the plankton floating in the sea depend on the supply of nutrients and the amount of sunlight penetrating the water (to allow photosynthesis to take place). The onset of ice activity on Antarctica had the effect of considerably extending the overall range of mineral nutrients and boosting their concentration levels in the surrounding seas. This dramatically increased the populations of various planktonic organisms living off such nutrient sources.

At the same time, erosion in the New Zealand area had worn down much of the land to below sea level so that shallow, largely sediment-free seas covered extensive areas of the Campbell Plateau and eastern New Zealand. In the New Zealand region,

OLIGOCENE IN NEW ZEALAND

Wearing-down of the ancestral New Zealand landmass reached a climax in the early Oligocene, when virtually the entire land was worn down to a very low level. The remaining remnants of land consisted of an elongated, narrow-gutted archipelago and a few scattered islands (Fig. 54). In some groups of animals and plants living in New Zealand at this time, isolation on islands may have stimulated development of many of the distinctive organisms seen today. For example, it is likely that some of the large native snails, first isolated at this time, evolved into species that today still retain their separate identities. Nonetheless, Oligocene times probably presented problems of survival for many of the other terrestrial plants and animals because the lowered topography of the land had reduced the number and variety of available terrestrial habitats. Widespread flooding of land by the sea further restricted habitats. Thus many extinctions of land organisms may have occurred at this time, thinning out still further the original plants and animals that had arrived from Gondwanaland in earlier Jurassic and Cretaceous times.

By contrast, conditions in the extensive nutrient-rich shallow shelf seas surrounding the New Zealand landmass favoured the development of cool-temperate marine life. The Oligocene seas of New Zealand became a major centre for evolution of penguins, whales, dolphins, and porpoises (as indicated by the little drawings on the map, Fig. 53). Although penguins had first appeared in the New Zealand region in the late Eocene (Fig. 51), a considerable expansion, both in numbers and variety occurred in the early and late Oligocene, and important fossils of these ages are known from South Canterbury and Otago.

Shallow seas of the ancient Tethys Ocean, lapping onto parts of southern Europe and Asia, initially served as a focal point for the evolution of the Cetacea (whales, dolphins, porpoises) in early and mid Eocene times. By late Eocene the focus had moved into the developing Pacific Ocean, because primitive toothed whales are known from the late Eocene of South Canterbury and Otago (New Zealand), Seymour Peninsula (West Antarctica), and Vancouver Island (Canada).

Ancestors of toothed whales, dolphins, and porpoises (Odontoceti) and baleen whales (Mysticeti) first appeared in the early and middle Oligocene of New Zealand. Development and expansion of these ancestral stocks, and also those of penguins, proceeded rapidly in the seas surrounding New Zealand. By late Oligocene times there was a great variety of early types of whales, dolphins, porpoises, and penguins in many areas, especially in the South Island. These New Zealand faunas had close similarities with Australia, South America, western North America, Japan, and Europe, showing that by late Oligocene times good circulation patterns had been developed in the major oceans.

The clear shallow temperate seas around the New Zealand landmass also provided optimum conditions for other forms of life, notably shellfish, echinoderms (sea eggs, sand dollars), moss-like bryozoa (lace coral) and planktonic organisms. As much of New Zealand was covered by sea during Oligocene time, marine deposits of this age, especially limestones, are well represented in the modern landscape and many of the deposits are particularly rich in a great variety of fossils.

Many of the Oligocene shellfish are newcomers, derived from Australian or Indo-Pacific sources. Such newcomers continued to arrive throughout the Oligocene, in spite of the fact that New Zealand was now surrounded on all sides by deep ocean. The marine creatures that arrived were all capable of being transported by oceanic currents as eggs, larvae, or adults. To such creatures deep ocean was, therefore, not a barrier in itself, but dispersion was dependent on factors such as sea water temperature and directions of prevailing winds and ocean currents. Ancestral marine stocks that had arrived in New Zealand in earlier times gave rise to a number of endemic forms (i.e., native to New Zealand) including the whelks *Buccinulum* and *Struthiolaria*.

Representative Oligocene shellfish are shown on the margin of Fig. 53 (clockwise from the top): *Maoricardium* and *Perna* from presumed Malayo-Pacific sources; *Mesopeplum* and *Bassina* from presumed Australian sources; *Buccinulum* and *Struthiolaria* from presumed endemic sources.

therefore, the cool nutrient-rich water from Antarctica was able to flow northwards and spread out across a large area of temperate shallow waters. Because of the virtual absence of erosion these waters had a high degree of clarity and, therefore, good light penetration, to allow maximum photo-synthesis (and hence growth) in the plankton.

All factors favoured substantial expansion of planktonic life and development of large populations of many other marine organisms either living off the plankton directly, or preying on plankton feeders. It is, therefore, not surprising that in the Oligocene the New Zealand region played a leading role in development of varieties of plankton feeders, notably Cetacea (whales, dolphins, and porpoises) and penguins.

OLIGOCENE LIFE

The Oligocene was a transitional period, when many of the Paleocene–Eocene plants and animals gradually died out or evolved into groups that eventually were to become members of the modern fauna and flora.

On land, the various mammal groups, including ancestral primates, continued to expand and diversify throughout Oligocene time. In the sea, differences appeared between Northern and Southern Hemi-spheres. In the Northern Hemisphere the effects of general cooling of the earth result-ing from the build-up of ice in Antarctica produced a marked (but temporary) slowing down of development of planktonic life. By contrast, in the Southern Hemisphere the advent of the circum-Antarctic current, bringing nutrient-rich Antarctic water into mid latitudes, stimulated rapid expansion of the plankton and evolution of plankton-feeding animals such as penguins, whales, porpoises, and dolphins.

Fig. 54 The gradual wearing-down and submergence of the ancestral New Zealand landmass created by the Rangitata Orogeny (140–110 million years ago) reached a climax in the Oligocene (37–25 million years ago), as depicted in the right hand diagram. During this period some two-thirds of the area of modern New Zealand was covered by sea.

In the Early Oligocene sea flooded in over the late Eocene coal swamps of Northland, Waikato, and Taranaki, extending eastwards at least as far as National Park. Deep seas covered the entire eastern flank of New Zealand, from East Cape to Southland. Sea extended from east to west across areas now largely occupied by the ranges of both North and South Islands. The South Island was virtually submerged, except for part of the old eroded plainland in Canterbury and Otago. Oligocene coal swamps formed in low-lying areas in Southland and Otago. The Fiordland region was largely submerged and a deep marine trough occupied the Waiau–Hollyford region. A chain of volcanic islands erupted in inland Canterbury and Marlborough. At East Cape and Oamaru, the volcanic activity that began in the late Eocene continued into the Oligocene.

Land is indicated by the coloured area, the Alpine Fault by a broken line, volcanoes by cone symbols, and coal swamps by a swamp symbol. Although the outline of modern New Zealand is shown on this diagram it is included for reference purposes only, as it did not physically exist as such at this time.

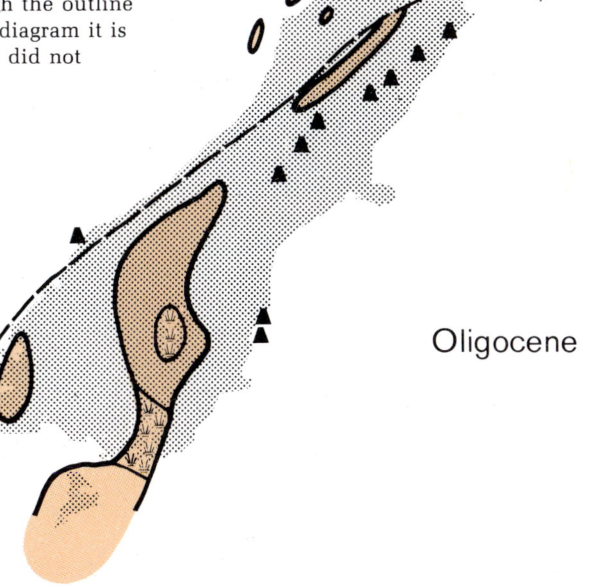

The steady northwards drift of New Zealand and Australia moved them into mid-latitude regions of the Southern Hemisphere. At the same time, sea-floor spreading movements in the South China Sea were rotating Malaysia, Indonesia, and the Philippine Islands in a southwards direction, gradually closing the gap between Indonesia and Papua New Guinea. The closer proximity of the Indo-Pacific land masses to each other meant that an increasing number of oceanic currents of tropical origin were able to reach the Australian and New Zealand coasts, bringing with them a great variety of tropical organisms capable of floating or swimming across open ocean. The south-ward spread of tropical organisms was also helped by a general improvement in climate that was occurring world wide at about the same time (early–middle Miocene).

Progressive narrowing of the oceanic gap between Papua New Guinea and Indonesia is reflected in the faunas that were able to bridge the gap at various times. In the Paleocene, flying animals such as birds and bats were able to bridge the gap by flying from Southeast Asia to Australia, and eventually on to New Zealand and New Caledonia (Fig. 49). Later, in the Eocene and Oligocene, a narrowing of the gap facilitated development of oceanic currents capable of carrying tropical marine organisms southwards over the same route. Then in the Miocene the intervening gap narrowed to such an extent, and had probably been at least partially bridged by volcanic archipelagoes, that land snakes (Fig. 55) were able to move into Papua New Guinea and eventually into Australia. (The two

countries were joined together until comparatively recent times.)

Between 80–60 million years ago the Tasman Sea had opened up, and ancestral New Zealand and New Caledonia had lost their land links to the remainder of eastern Gondwanaland. Snakes, therefore, were unable to reach either New Zealand or New Caledonia. Although there is some evidence that ancestral snakes may have reached South America earlier in the Cenozoic, or perhaps even at the end of the Cretaceous, it is probable that a land route to Australia via eastern Antarctica had already been disrupted by sea-floor spreading between Patagonia and the Antarctic Peninsula. Even if such a route was available it is likely that it was cool-temperate in character and therefore impassable to cold-blooded creatures such as snakes.

Continuation of the movement of Australia and New Zealand northward had the effect of opening up even more of an oceanic gap around Antarctica. The circum-Antarctic current, first established in the Oligocene, became a major element in the Southern Ocean in the Miocene. Such a current system, together with associated winds, was capable of transporting marine creatures able to swim or float as adults, juveniles or larvae. In this way new "southern" distrib-utions of marine organisms were achieved.

Some marine animals of tropical or sub-tropical origin, moving southwards in currents flowing from the Indo-Pacific region, were able to link up with the circum-Antarctic current and by this means

became widely distributed around southern as well as tropical lands. Examples of such distributions occur in echinoderms, gastropods, and bivalves.

Cooling of the Antarctic continent, and the spread of cool Antarctic water by means of the circum-Antarctic current, led to development in the early Miocene of major temperature differences between polar and tropical waters, similar to that of today. These temperature differences were expressed not as a continuous gradation from pole to tropics but rather as a series of belts of oceanic waters of differing charac-teristics, separated by boundary zones or "convergences" (Antarctic Convergence, Subtropical Convergence). While these belts have largely retained their identity from early Miocene times through to the present day, their position with respect to latitude has fluctuated in response to episodes of global warming and cooling.

Differentiation of separate Antarctic, sub-Antarctic, and temperate water masses in the Southern Ocean, distinct from those of the tropics, was matched by development of appropriate climatic zones and meteor-ological conditions and during Miocene times southern oceanic climates in general began to approach those of today. None-theless, the coastline of Antarctica in the Miocene still remained largely free of ice, except for the terminii of large glaciers flowing from inland regions. In sheltered areas the coasts were probably fringed with luxuriant growths of kelp, like those of the sub-Antarctic islands today.

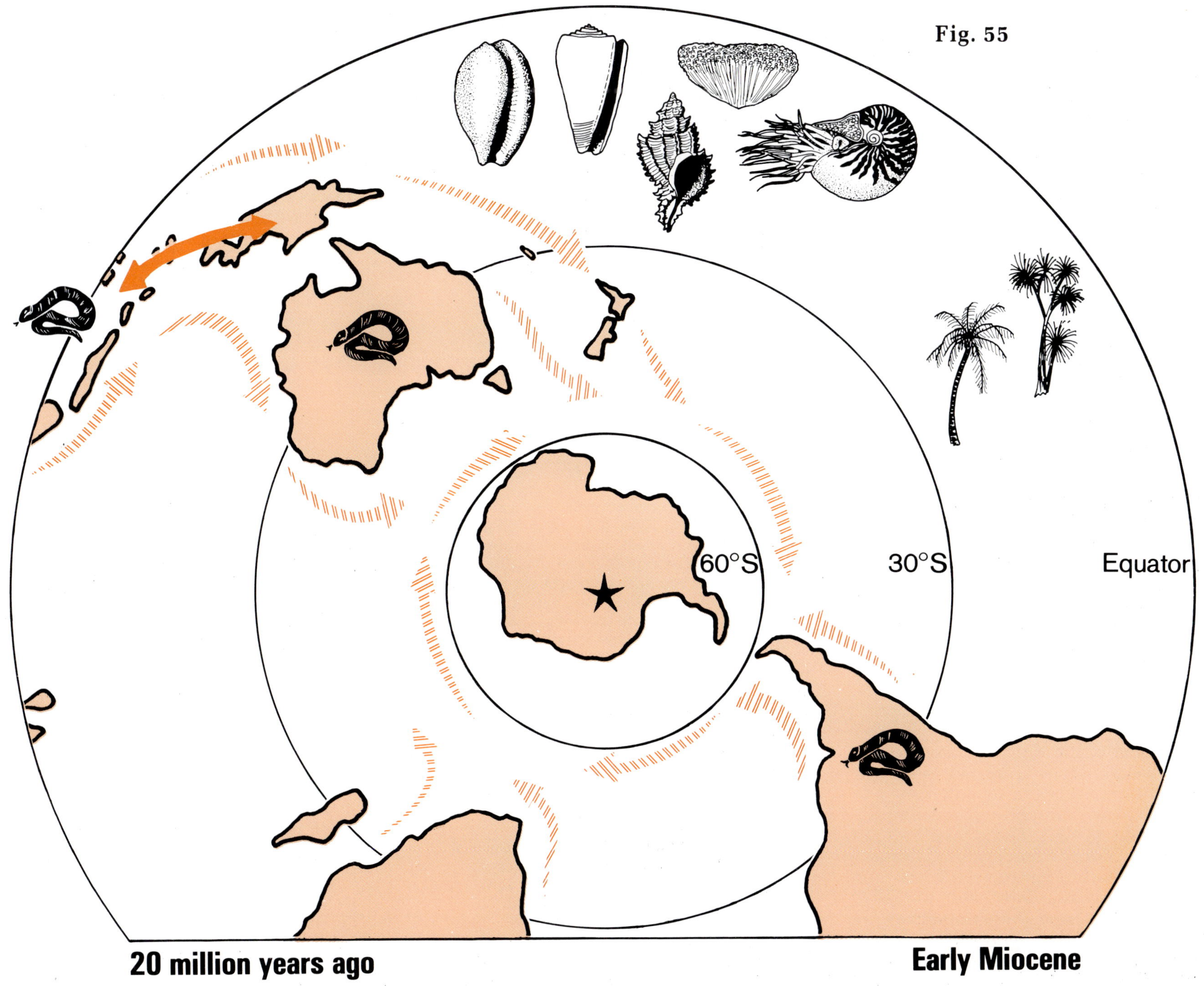
Fig. 55
60°S
30°S
Equator
20 million years ago
Early Miocene

MIOCENE LIFE

The Miocene is characterised by the appearance and radiation, or continued development, of animals and plants ancestral to modern forms. Development of mammals reached a maximum in the Miocene. Groups to diversify substantially in the Miocene included ancestors of elephants, deer, pigs, tapirs, and rhinoceros. Monkeys appeared as a distinct group in the middle and late Miocene.

Changes in climate favoured the spread of grassland in many areas of the world, and a corresponding reduction in woodland. This caused a great radiation of grazing herbivores ancestral to modern antelopes, cattle, horses, sheep, and goats. The tough grasses, rich in abrasive silica minerals, presented a problem for those animals relying on leaf protein as a food substance. This is because mammals have a limited number of teeth and if all the teeth are worn away the animal will starve to death. Increase in grassland therefore favoured those groups of grazing herbivores with flat grinding teeth, continually growing throughout life, whereas some other groups of plant-eating mammals without such teeth became extinct.

EMERGENCE OF MAN

Towards the end of the Miocene (14–10 million years ago), fossils appeared in Pakistan and China that are regarded as ancestral to modern humans and great apes (chimpanzee and gorilla). These fossils were probably rather like the modern orang-utan, gibbon, and gorilla in appearance. It is thought the lineages that later were to lead to man on one hand, and the chimpanzee and gorilla on the other, separated from one another sometime between 10 and 5 million years ago. Evidence from skeletons and from footprints found in Ethiopia and Tanzania indicate that ancestral humans were already walking upright and had an erect posture by 3–3.5 million years ago. These ancestral humans, called *Australopithecus*, were slightly more than one metre in height. They had a small brain with a capacity of 500 cm³, comparable to that of the modern gorilla. Their head displayed a mixture of ape and human characteristics: a low forehead and a long ape-like face, but with teeth proportioned like those of modern man.

Ancestral humans inhabited the newly-developed grasslands of Africa, Europe, and Eurasia. Tool-making humans, (*Homo habilis*, meaning "handyman"), appeared about 2 million years ago. They had teeth more man-like than those of *Australopithecus*, and a larger brain, with a capacity of 700 cm³. *Homo habilis* was succeeded about 1.5 million years ago by *Homo erectus* ("erect man"). Compared to modern man, *Homo erectus* had a flat braincase, heavy brow ridges and a thick robust jaw. The brain capacity of *Homo erectus* was 900–1000 cm³, compared with that of modern man, *Homo sapiens*, of 1400 cm³. *Homo erectus* made a great variety of stone tools and was the first known user of fire. Modern man, *Homo sapiens*, appeared about 500 000 years ago.

Fig. 56 In the early Miocene (25–18 million years ago), as depicted in the left hand diagram, the long episode of progressive erosion and submergence of the ancient New Zealand landmass was brought to a close. The earth movements that eventually were to shape modern New Zealand caused the sea to withdraw from many areas that had been inundated in Eocene and Oligocene times. New land began to appear in both North and South Islands (compare with Fig. 52, 54). Swampy river deltas in the north Taranaki area laid down deposits that were later to become the Mokau coals. A rising ridge in Northland and Auckland was flanked by a chain of andesite volcanoes and the first volcanic eruptions began in the Coromandel Peninsula. Apart from volcanoes, however, the early Miocene land was not yet mountainous. During the early Miocene warm water plants and animals colonised most parts of New Zealand, and coral reefs grew around the coasts as far south as East Cape.

In the late Miocene (12–5 million years ago), as depicted in the right hand diagram, more and more land began to rise up out of the sea. For the first time since the earliest Cenozoic, high hills, verging on mountains, began to make their appearance in Marlborough, southern Nelson, southern Wellington, and southern Hawkes Bay. The system of faults associated with the Alpine Fault today (Fig. 7) started to become active at about this time. Fault-bounded blocks of land began to rise independently of other areas, foreshadowing development of the modern main ranges of both North and South Islands. Faults are indicated by broken lines on the map. Swamps formed in low-lying areas in Southland, Central Otago, and Murchison, giving rise to lignite and coal deposits. An extensive chain of andesite volcanoes was active in Coromandel. The gold, silver, and other ores of the region are related to hot mineralised fluids given off at this time. A chain of volcanoes was also active off the north Taranaki coast. Volcanic activity started at Lyttelton and the Otago Peninsula, and continued into the Pliocene (see Fig. 60).

Land is indicated by the coloured area, volcanoes by cone symbols, and coal swamps by a swamp symbol. Although modern New Zealand is shown in outline, it is for reference only, as it did not physically exist at this time.

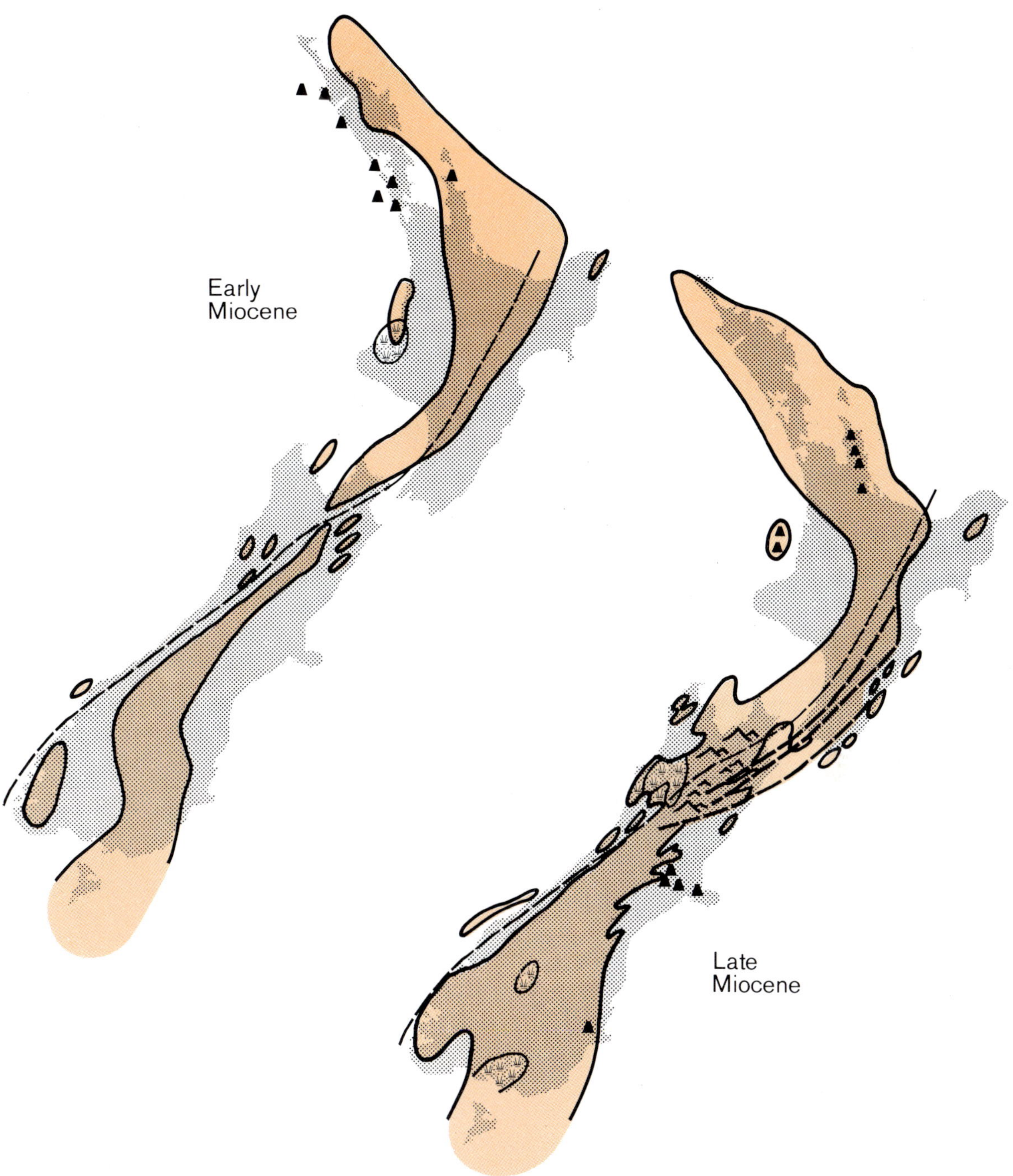

The early Miocene marked the high point of Tertiary climate. Thereafter a period of cooling ensued and middle and late Miocene times saw a gradual disappearance of many of the tropical organisms that had previously migrated to New Zealand. Sea water temperatures around New Zealand coasts however were still warmer than those of the present day.

As climate cooled, the Antarctic ice sheets grew in size and the geography of the Southern Hemisphere began to approximate that of today. The meteorological and open-sea conditions surrounding Antarctica provided further enhancement of the circum-Antarctic current and, along with other associated wind and oceanic current systems, it became the dominant mode of transport of organisms in the Southern Hemisphere (indicated diagrammatically by the arrows in Fig. 57). New Zealand marine organisms that were probably originally distributed in this manner in the late Miocene are illustrated (clockwise from the top): echinoderms such as *Pseudechinus*, shells such as *Lamprodomina* (olive shell), *Aeneator*, and *Atrina* (horse mussel).

The influence of the circum-Antarctic current system on the seas surrounding Antarctica, first felt in the late Oligocene, became greatly enhanced in the Miocene by the progressively widening gaps between Antarctica, South America, and Australia. These widening gaps allowed increasing quantities of water to be involved in the current movements. Further enhancement also came from climatic deterioration in middle and late Miocene times. The resulting build-up of ice sheets on Antarctica led to generation of Antarctic-centred weather patterns.

Establishment of the oceanic circum-Antarctic current brought into being a range of associated meteorological phenomena that have a decisive influence on Southern Hemisphere weather. The most notable of these phenomena is the West Wind Drift— prevailing westerly winds that encircle the globe at latitudes between 40°S and 60°S (see Fig. 64). They give rise to the "Roaring Forties", "Furious Fifties", and "Screaming Sixties". These winds are so powerful and constant that they can drive surface water along at rates matching the deeper flowing circum-Antarctic current. The transporting efficiency of such a wind and current system is so high that once established it acted as a dispersal mechanism by which many animals and plants crossed the Southern Ocean after Antarctica ceased to be available as a stepping stone.

Thus from the Miocene onwards New Zealand began to gain many new southern colonists. These differed markedly from those it had received in the Cretaceous. The southern plants and animals of the Cretaceous had travelled across and around the then closely connected southern lands and were all of types that needed either continuous land, continuous shallow water, or closely adjoining islands for dispersal. The riders of the West Wind Drift, on the other hand, were all of types capable of crossing major water barriers as eggs, larvae, or adults. As the distribution of land and sea in the Southern Hemisphere from Miocene times onward was virtually as it is now, it is evident that these organisms were capable of surviving long-distance dispersal over oceanic gaps of up to 6000 km.

Possible mechanisms for such long-distance travel include transportation by birds, ocean currents, and winds. Many birds, such as giant petrels, shearwaters, and albatrosses, migrate widely in southern latitudes, often assisted by the West Wind Drift. They may carry seeds encaked in mud around their legs, hooked seeds attached to their feathers, or fleshy seeds in their stomachs. Ocean currents, driven by the West Wind Drift, can carry floating material between the southern lands. Floating or swimming animals, and plants with coastal habitats and buoyant fruits, are prime candidates for this type of dispersal and it is believed that plants such as *Hebe* and kowhai (*Sophora*) migrated between New Zealand, southern Chile, and Gough Island in this way.

The high winds of the "Roaring Forties" may, in their own right, distribute plants and animals. Many plants have light seeds or spores, or plumed seeds, and some spiders, for example, create web parachutes by means of which they can "balloon" in the wind.

Although Australia has always made contributions to New Zealand's flora and fauna throughout most epochs of geological history, trans-Tasman migration was greatly strengthened by the advent of the West Wind Drift. From Miocene times onwards, therefore, New Zealand received from across the sea numerous Australian plant and animal migrants, both marine and terrestrial.

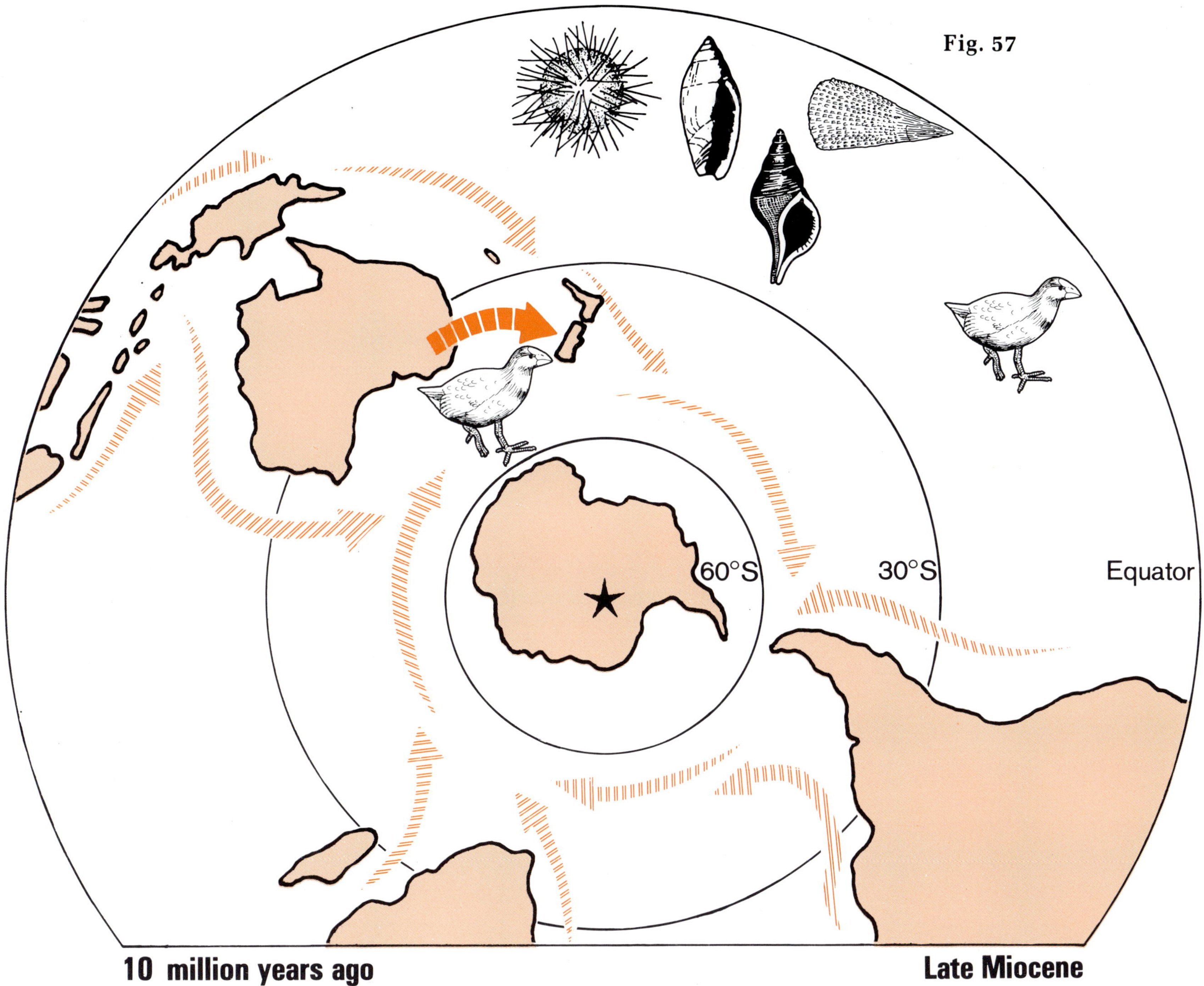

Fig. 57
60°S
30°S
Equator
10 million years ago
Late Miocene

Land birds were probably also notable riders of the west wind. The late Cenozoic saw the arrival of many land birds of Australian origin, travelling via storm-generated westerly gales. As a result, the modern New Zealand bird fauna has a strong Australian aspect. Notable exceptions are the moa and kiwi dating back to late Jurassic and early Cretaceous times when southern land links were evident. Also endemic are the families of New Zealand wrens, thrushes, and wattle-birds, probably introduced as wind-assisted migrants in early Cenozoic times, before distances became too great across the widening oceans around New Zealand.

The continuing influx of storm-blown Australian land birds into New Zealand throughout late Cenozoic times has given rise to examples of multiple colonisation. The takahe, for example, as illustrated on the map, is derived from an old (Miocene–Pliocene?) migration across the Tasman and has diverged considerably from the original Australian stock, whereas the pukeko is from a very recent (Holocene?) migration and is indistinguishable from Australian forms.

Older land-bird migrants to New Zealand became modified in various ways, notably flightlessness and giantism. In most environmental situations flight is of prime importance, allowing birds to escape predators, and to nest in the comparative safety of trees. The necessity of flight has tended to restrict the size of birds, as flight is easier for small light birds. However the New Zealand environment, without flesh-eating predators, tended to encourage flightlessness and ground-nesting habits. Loss of flight in turn paved the way to giantism, and as the birds became larger other changes occurred: the feathers grew heavier and sparser, the legs stouter and shorter, and the brain became relatively small in relation to the beak, jaws, and cheek muscles. This was a response to the adoption of a grazing habit, like grazing animals elsewhere in the world. The takahe, kakapo, and extinct goose (*Cnemiornis*) are examples of late Cenozoic land bird migrants that became large and flightless, whereas on the other hand the New Zealand

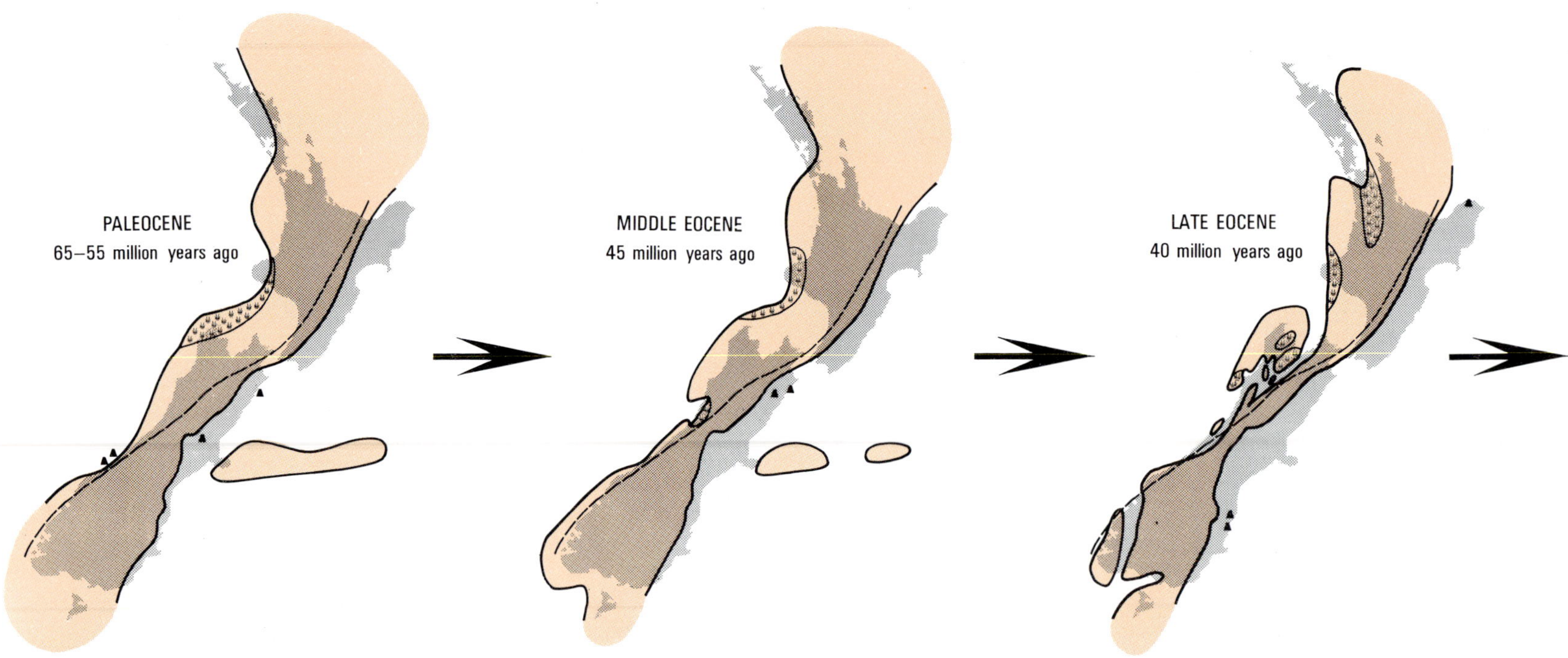

robins, for example, became larger without completely losing the power of flight.

Trans-Tasman migration of Australian land birds continues today and colonists arriving in the past century include the spur-wing plover, black-fronted dotterel, white-faced heron, Australian coot, royal spoonbill, grey teal, welcome swallow, and wax eye. Other would-be colonists have lingered but not survived, e.g. avocet, little bittern and white-eyed duck. Many Australian insects also arrive in the aftermath of westerly gales, but few survive to colonise, although the monarch butterfly is a notable survivor. Australian fruit bats have also been blown across the Tasman, but have not formed colonies.

The trans-Tasman migrants of late Cenozoic times arrived in New Zealand to find a land with a wide range of temperate maritime climates, a diversified landscape ranging from high country to alluvial plains, and a broad spectrum of ecological habitats.

The worn-down and largely submerged remnants of the late Mesozoic and early Cenozoic ancestral New Zealand landmass had in the early Miocene given way to a long narrow archipelago (Fig. 58). Although many earth movements were underway, there was little if any mountainous terrain. By late Miocene times, however, large areas of rugged terrain had been uplifted, and a landscape similar to that of modern New Zealand had started to take shape.

Fig. 58 New Zealand's changing geography. The ancient New Zealand landmass created by earth movements (140–120 million years ago) was gradually reduced by erosion and eaten into by the sea until, in the Oligocene (37–25 million years ago), the land was virtually entirely submerged except for a few islands and archipelagos. Earth movements then began, initially in the early Miocene, but gaining momentum in the late Miocene and succeeding geological periods. These movements eventually created the modern New Zealand landmass. The diagrams summarise New Zealand's changing geography over the period between 65 and 6 million years ago. Land is indicated by the coloured area, the position of the coastline by a solid line, the Alpine Fault system by a dashed line, volcanoes are indicated by solid triangles, and areas of coal formation by a swamp pattern. The present shape of New Zealand is indicated by the shaded outline.

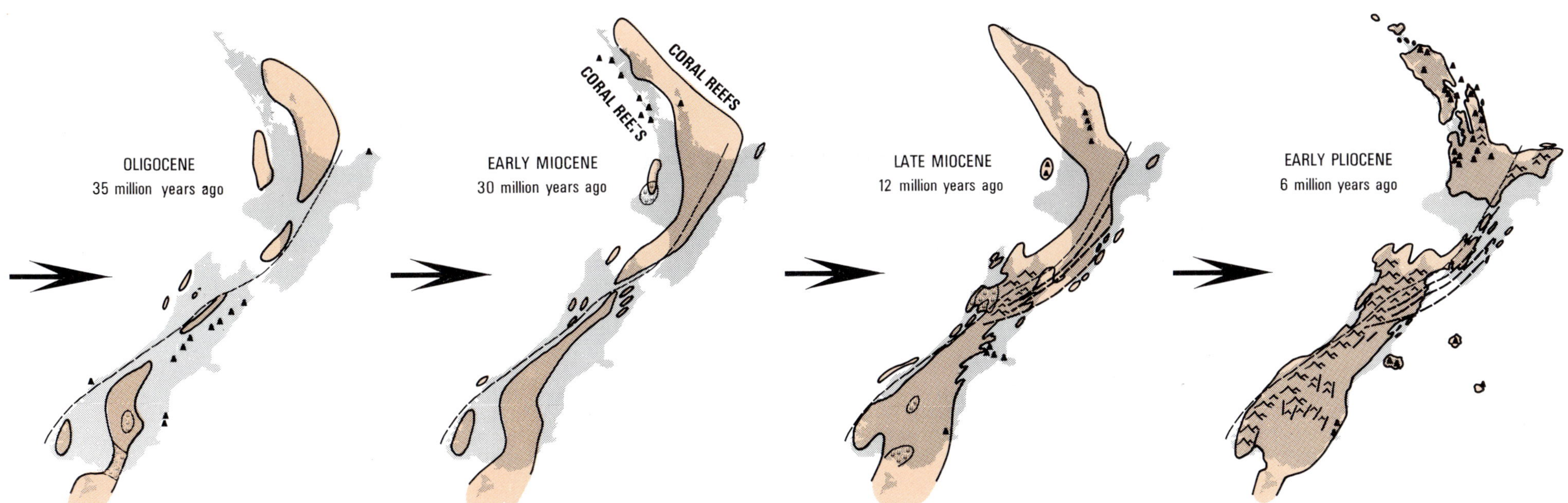

Despite its short length (3 million years), the Pliocene was a period of marked and comparatively rapid change. It was the last geological epoch before the world became very much as we know it today. The beginnings of a great number of changes occurred in the Pliocene: world geography changed, mountain ranges were formed, climate changed, animals and plants changed. These changes continued into the Pleistocene, and eventually shaped the Earth we now know. Towards the end of Pliocene time the outlines of many of the continents became very similar to those of today, except that sea still covered the edges of the land in many countries, notably North and South America, Italy, the Middle East, Japan, Indonesia, and New Zealand.

In the Southwest Pacific, Australia and New Zealand continued to travel north in tandem into progressively warmer climatic zones. A reflection of this northwards movement was the apparent ease with which warm-temperate and subtropical marine organisms were able to enter both countries (although such migrations also depended on appropriate oceanic currents being available). In both countries, however, migrations from northern sources were also accompanied by influxes of southern migrants transported by the steadily strengthening systems of circum-Antarctic winds and oceanic currents (as indicated by the arrows, Fig. 59). For this reason, the patterns of floral and faunal migrations are a complex mosaic and not necessarily a simple expression of the steady northwards movement of both countries.

A timetable for northward movement of Australia (and by inference, New Zealand) can be obtained from traces of ancient dust storms. Material from dust storms originating in Australia, has settled into the Tasman Sea to form clay layers on the sea bed. Study of these layers has enabled geologists to track Australia's entry into the latitudes of hot arid equatorial climate and, by extrapolation, New Zealand's movement into warm-temperate and subtropical latitudes. Dating and analysis of the clay layers indicates that deserts began to form in northern Australia about 14 million years ago (middle Miocene). Throughout the late Miocene, as Australia gradually drifted northwards, deserts extended southwards. By early Pliocene times they had an extent similar to that of the present day. According to this evidence, by the early Pliocene Australia and New Zealand had reached latitudes similar to those they occupy today.

Pliocene times saw the culmination of a series of step-like events leading up to the great cold periods that affected many areas of the world in the Pleistocene (Ice Age). Until late Miocene times many of the crucial events took place in the Southern Hemisphere. The first step began in the Eocene with the southward movement of Antarctica, resulting in the accumulation of snow and ice. At first it gathered on the high country of inland Antarctica (early–middle Eocene), but later extended outwards as valley glaciers to reach the Antarctic coastline (latest Eocene–earliest Oligocene). This was followed in late Oligocene by movement apart of the various

southern landmasses, creating open oceanic areas for the development of the circum-Antarctic systems of winds and ocean currents. This effectively isolated the Antarctic continent from the influence of warmer waters to the north.

The meteorological and oceanic conditions thus established led to development of ice sheets in eastern Antarctica in the middle Miocene (11–12 million years ago) and western Antarctica in the latest Miocene (5 million years ago). A massive increase in the size of all the Antarctic ice sheets occurred in the early Pliocene, 4.3 million years ago.

In the late Miocene and Pliocene these events in the Southern Hemisphere were augmented by events in the Northern Hemisphere. Opening up of the North Atlantic provided a high-latitude source of cold water, flowing southwards into the North Atlantic. Newly formed mountain ranges appeared in many of the Northern Hemisphere lands, and began to exert a considerable influence on the climate of neighbouring regions.

As Africa and the Middle East moved northwards into contact with the southern edge of Europe and Eurasia, the ancient Tethys Ocean was broken up into a series of separate seas, of which the Mediterranean, Black, Caspian, and Aral seas are the remnants today. The gradual disappearance of the Tethys removed the moderating climatic influence of open oceanic conditions from large areas of southern Europe and the Middle East, and undoubtedly

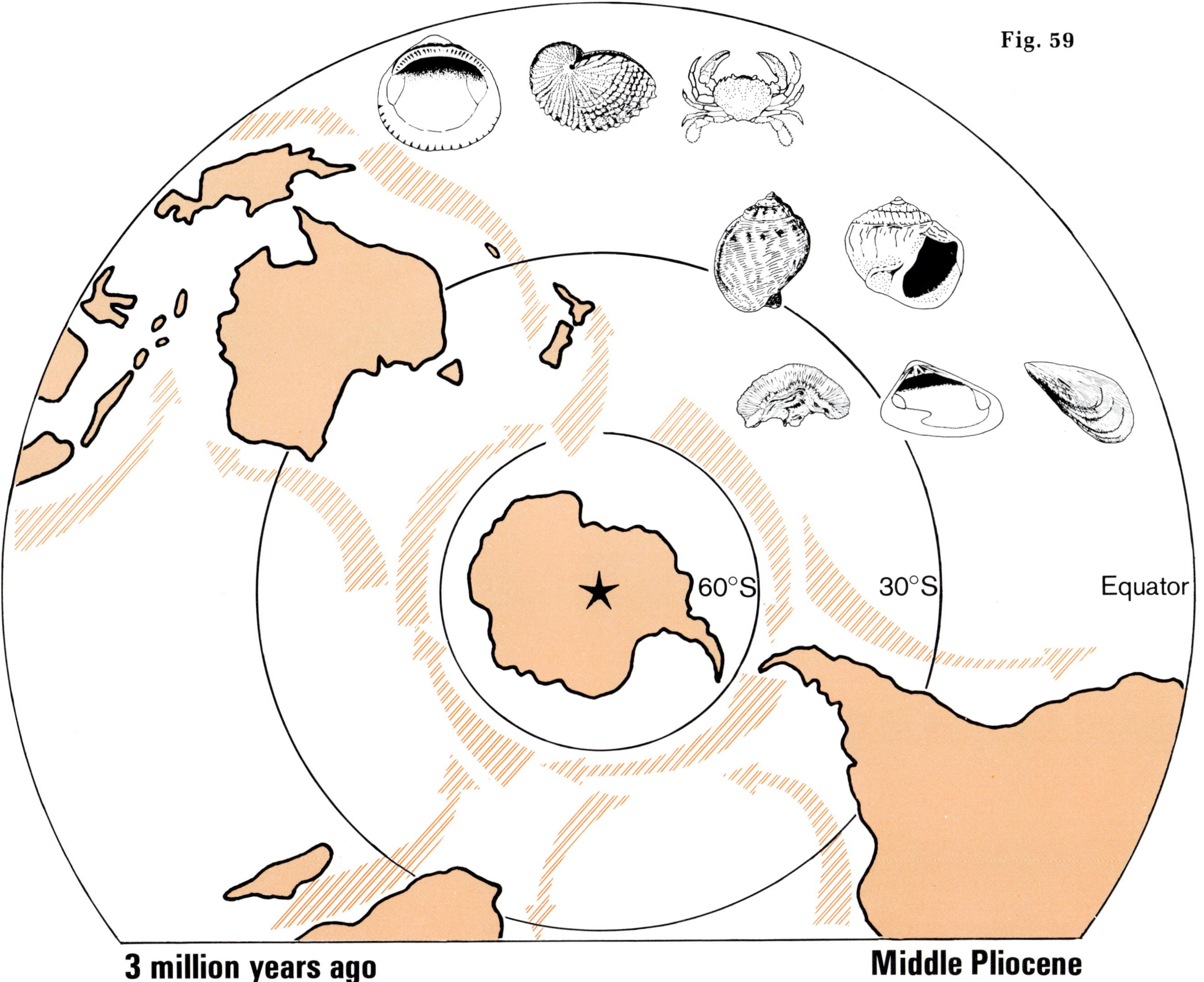

Fig. 59
60°S
30°S
Equator
3 million years ago
Middle Pliocene

resulted in a re-arrangement of oceanic circulation in low latitudes.

Comparable changes in climatic and oceanic circulation were also caused by geographic changes in the Americas. In the late Plio-cene (3.5–3 million years ago), earth movements in Central America formed the Isthmus of Panama, stopping the free inter-change of water between the Atlantic and Pacific Oceans. This interrupted the circum-equatorial current system and diverted increased volumes of warm water into high latitudes, stimulating an increase in evapo-ration, increased falls of snow and ice over northern Canada and Greenland, and later development of glaciers and ice caps. The cooling of the Northern Hemisphere oceans that followed favoured the build-up of even more ice. The first large accumulations of ice on northern lands became apparent 3 million years ago, and the first major expansion of sea ice occurred in the North Atlantic 2.4 million years ago.

PLIOCENE LIFE

The climatic and geographic changes that occurred during Pliocene times had a major effect on the global distribution of plants and animals. Development of polar glaci-ation in both hemispheres led to the appearance of various cold-adapted polar faunas and floras. The deteriorating Pliocene climate wiped out many of the species that had lived in the more equable climates of the Miocene. For example, many of the unusual Miocene mammals died out in the Pliocene. However, other animal groups either expanded into the ecological niches thus vacated or, continuing their evolu-tionary paths, gave rise to entirely new forms. Ancestral elephants, rhinoceroses, and giraffes first appeared in the early Plio-

cene and the first mastodons and mammoths in the late Pliocene and early Pleistocene. *Australopithecus*, man's earliest definite ancestor, appeared in middle Pliocene times, 3.4–3.0 million years ago.

The Pliocene marked the climax of evolution of the horse family. During the Miocene, two horse groups had evolved in North America: browsing woodland horses and grazing plains-living horses. By the end of the Miocene, the woodland horses had declined in numbers and later, in the middle Pliocene, became extinct. The plains horses took advantage of the abundance of newly evolved grasses and became very numerous and diversified. In the early Pliocene they crossed Bering Strait into Siberia, either by island-hopping, or moving across land exposed during times of low sea level. From Siberia ancestral horses spread across the grasslands of the Old World.

In the same way as ancestral horses took advantage of land links in the north to cross Bering Strait and gain entry to the Old World, many other animals at the southern tip of North America were using the newly created Isthmus of Panama to gain entry to South America. Uplift of land to form the Isthmus of Panama in the late Pliocene (3.5–3 million years ago), brought to an end the long-continued isolation of South America and in doing so sounded the death-knell for one of the most unusual assemblages of mammals the earth has ever seen.

South America and North America had been separated since at least late Jurassic times, 140 million years ago. Any links had been very temporary, via circuitous routes by way of Europe and Africa, or by island-hopping across short-lived volcanic archipe-lagoes. Until the late Pliocene, no land mammals were shared by the two conti-nents. Both had a range of mammals that

although occupying the same ecological niches, and often showing at least some superficial similarities, had been derived from vastly different stocks. In South America they were derived from marsupials, and unusual ungulates (hoofed animals), whereas in North America they were derived entirely from placental stocks.

The South American mammal fauna developed an incredible variety of forms, many of enormous size, and most now extinct. Carnivorous marsupials included powerful bear-sized animals, as well as forms similar to pumas, coyotes, and foxes. But the most remarkable was the tiger-sized sabre-toothed marsupial cat. The rather strange hoofed mammals, totally unrelated to any we know today, resembled camels, hippos, rhinoceroses, horses, and elephants. Other South American placental mammals are the edentates (meaning toothless). They include herbivorous and insectivorous anteaters, tree sloths, ground sloths, and armadillos with their strange extinct giant cousins the glyptodonts.

By Pliocene times there were 27 families of mammals in North America and 29 in South America, and not one family in common. Then, at the end of the Pliocene, the Isthmus of Panama came into being. The two faunas began to mingle, and 22 mammal families came to be shared by North and South America. Not all the mammals in North and South America crossed over the land bridge. Some were cold or temperate climate animals which could not tolerate the tropical temperatures of Central America. Nevertheless, mass migrations of animals occurred across the isthmus in both direc-tions. The animals which headed north from South America included the now extinct ground sloths, but also armadillos, porcu-pines, and opossums still living today in parts of North America.

The animals which went south from North America were more numerous, and far more successful. They included dogs, llamas, horses, tapirs, rabbits, elephants, and sabre-tooth cats. Their success resulted in the extinction of many of their South American counter-parts. Eventually, even some of the immigrants themselves (sabre-tooths, elephants, and horses) became extinct. Horses eventually disappeared from both North and South America and were re-introduced from Europe by the Spanish only a few centuries ago.

Another unusual geographic event, with major consequences for the life of the time, was the drying-up of the Mediterranean Sea. The closing of the Tethys Ocean pinched-off the western end of the Mediterranean in the latest Miocene (5 million years ago), and isolated the Mediterranean waters from the Atlantic. The eastern end of the Mediterranean was also closed, creating the land areas of the Middle East. These land connections enabled some Eurasian mammals to move south into Africa (rhinoceros, antelope, jackal, lion, hare) and some African mammals to move north into Eurasia (monkey, porcupine, elephant).

Large areas of the land-locked Mediterranean Sea dried up through evaporation, although a few saline lakes were probably left in the lowest parts. Thick deposits of various salt minerals accumulated on the dry sea floor. The great increase in salinity and the shrinking of the sea killed off most of the marine organisms. The drying-up of the Mediterranean took place over about 500 000 years or more. In the earliest Pliocene, however, the waters of the Atlantic Ocean broke through the Straits of Gibraltar, and the Mediterranean was gradually filled again and re-populated with marine creatures.

Fig. 60 In the early Pliocene (5–3 million years ago), earth movements became even more intense and the geography of modern New Zealand started to take shape. The faults of the Alpine Fault System (cf. Fig. 7) were all active (indicated by lines), pushing up blocks of land that were to become today's main ranges. Sea occupied the Taranaki–Wanganui, Gisborne–Hawkes Bay, and southern Wairarapa areas. Deposits of Pliocene marine rocks were laid down in coastal areas of Westland, Canterbury, and Marlborough. The sea flooded in over some areas that had been land in the Miocene (cf. Fig. 56), notably the Manukau Harbour–lower Waikato area and the southern third of the North Island. Marine straits crossed the rising main ranges at the Manawatu Gorge and Kuripapango (inland Hawkes Bay). Such inundation was only temporary, however, as later in the Pliocene, and in the Pleistocene, continuing earth movements sent the sea into retreat. Volcanic activity was widespread in Northland, Coromandel Peninsula, and the Waikato (indicated by cone symbols). Twin basalt volcanoes were built on the site of Banks Peninsula, the eroded craters of which are now occupied by Lyttelton and Akaroa harbours. Port Chalmers occupies the eroded crater of another basalt volcano erupted on the site of modern Otago Peninsula.

Land is indicated by the coloured area. The outline of modern New Zealand is shown for reference, but it must be remembered that at this time it did not physically exist.

PLIOCENE IN NEW ZEALAND

During the Pliocene there were substantial areas of land in the New Zealand region (see Fig. 58), but records of land life are very sparse. Although some indication of the nature of the vegetation is obtained from fossil pollen and spores (preserved in lignites and marine sediments), there are no indications of the accompanying terrestrial animal life, apart from occurrences of fossil penguins and a bony-toothed seabird, *Pseudodontornis* (Fig. 61), preserved in marine sediments at Motunau, North Canterbury. As with virtually all the New Zealand fossil record (except for the late Pleistocene), details of Pliocene life are largely obtained from marine sediments preserved in areas that were flooded by Pliocene seas: the edges of the West Coast, Canterbury and Marlborough, Wairarapa, Taranaki, Wanganui, Hawkes Bay, lower Waitako, and Auckland (Fig. 60).

The marine fossils, as well as pollen and spores, provide evidence that in the Pliocene climate was generally becoming cooler in New Zealand, compared with that of the preceding Miocene. However, the sea water temperatures were probably nowhere cooler than those now found around Northland. The cooling waters around New Zealand quickened the northward retreat of warm-water species that had started in the late Miocene. Many such species were replaced by forms with southern distributions. Many of these newcomers arrived via the winds and ocean currents of the circum-Antarctic westerly system.

Creation of a fairly continuous NE–SW landmass, and elevation of rugged mountainous terrain to form a "backbone" to the land (Fig. 60), accentuated climatic differences between east and west coasts. The west coast tended to be influenced by warm water carried southwards from Australia by the East Australian Current, whereas the east coast was influenced by cold sub-Antarctic water. Because of this geographic situation, northern groups continued to arrive throughout Pliocene times in spite of the fact that many of the warm-water groups that had originally come to New Zealand in Miocene times became progressively extinct at the same time.

Fossils representing the various groups that migrated to New Zealand in the Pliocene are illustrated on Fig. 59. They are (clockwise): immigrants from Malayo-Pacific sources represented by the bivalve dog cockle *Glycymerula*, the paper Nautilus *Argonauta*, and the swimming or paddle crab *Ovalipes* (upper row); from Indian Ocean and Australian sources by the helmet whelk *Semicassis* and the mud whelk *Amphibola* (middle row); from southerly sources via the circum-Antarctic current system the temperate coral *Flabellum*, the pipi *Paphies*, and the ribbed mussel *Aulacomya* (lower row).

Entry of immigrant marine organisms to New Zealand fluctuated throughout Pliocene time, and a number of advances and retreats of cool-water forms occurred. These are interpreted as a response to climatic fluctuations preceding onset of the Ice Age. At first, these fluctuations were probably comparatively minor phases of cooling, but towards the end of the Pliocene they had apparently reached the proportions of small glaciations. In the late Pliocene and early Pleistocene (2.2 million years to 850 000 years ago), some 12 glacial events of short duration (about 40 000 years) have been documented in the New Zealand fossil record. The onset of the first of these glacial events coincides with the northwards migration of the subantarctic scallop *Chlamys delicatula* into Wairarapa and Hawkes Bay, although it did not attain its maximum abundance and northernmost extent until early in the Pleistocene.

The Pliocene seas covering those areas of New Zealand still submerged at that time (Fig. 60) laid down a great variety of sediments. Included amongst these were some very distinctive limestones, laid down in shallow water and composed predominantly of billions upon billions of barnacle plates, (but also containing oysters, scallops, and bryozoa). These limestones now form picturesque scarps and bluffs running across the country in coastal and inland Hawkes Bay and northern Wairarapa: Te Mata Peak and the bluffs extending southwards along the adjacent Tuki Tuki Valley are notable examples.

Another characteristic rock laid down during the Pliocene, but in deeper water, was a fine-grained, firm, pale bluish grey siltstone, commonly called "papa". Such papa rocks were deposited at various times during the Cenozoic throughout New Zealand so they vary in age from area to area. Fossils are sparse, and many of those present indicate that the papa was deposited well off-shore on the continental shelf, in depths of 100 or 200 m.

During the Pliocene a seaway stretched from southern Taranaki and Wanganui, across the present site of the Ruahine Range (extending from the Manawatu Gorge to the Napier–Taupo Road), and out to deeper water covering Hawkes Bay and adjacent parts of the modern continental shelf (Fig. 60). Sequences of marine limestone, sandstone, and siltstone were laid down on the edges of the seaway. The barnacle limestones of Hawkes Bay and northern Wairarapa were formed from enormous barnacle banks living along the shallow margins of the seaway. The barnacles thrived in those areas where strong tidal currents had removed mud and sand and scoured a clean surface on underlying older hard rocks. Maintenance of such conditions over a considerable period of time enabled the barnacle banks to both extend laterally into deeper water and to build up great thicknesses, often in response to rising sea level or subsidence of the seafloor. As a result, massive accumulations of barnacle plates built up that were geographically widespread as well as very thick.

Between 3 and 4 million years ago the sea bed started to rise slowly as new land began to form, especially along the major mountain ranges. This caused the water to shoal, the incoming sediments to become coarser (as the coast was closer) and shells to be more plentiful. Sandstones, coarse limestones, and shell grits were laid down in many areas. However, over the last 1 million years, as the Ruahine Range began to rise out of the sea and the old hard rocks in its core began to erode, massive conglomerate deposits of pebbles, gravel, and rock chips were formed.

Fig. 61 The Pliocene rocks of Motunau, near Christchurch, have yielded a jaw bone of a rather unusual bird, called *Pseudodontornis* or "false-toothed bird". These birds were among the largest flying birds and, with a wingspan of 4.5 m, they were well adapted for soaring. They were rather like enormous gannets, to which they were related, and they could swallow kahawai-sized fish by hinging their lower jaw. As a group, the false-toothed birds first appeared in the Eocene, but became extinct at the end of the Pliocene. All of the false-toothed birds had bony tooth-like projections along the sides of the bill. Although such "teeth" looked like true reptilian teeth, and probably functioned as such, they were simple bony continuations of the jawbone and lacked the enamel and dentine characteristic of real teeth.

PLEISTOCENE & HOLOCENE

(2 million years, to present day)   (Fig. 62–66)

The ice-free conditions in the early Cenozoic changed to widespread and numerous glaciations (about one every 100 000 years), in the Pleistocene (Ice Age). The transformation was brought about by: the positioning of Antarctica across the South Pole; movement apart of continents (enabling establishment of the circum-Antarctic systems of winds and currents 34–24 million years ago); development of permanent Antarctic ice sheets (12–5 million years ago); and the onset of major Northern Hemisphere glaciation in late Pliocene (3–2.4 million years ago).

It is difficult to pinpoint exactly the start of the Ice Age in all parts of the world, as it probably varied considerably with local conditions. In New Zealand the first signs of marked cooling were in the latest Pliocene (2.2 million years ago), when the first of a series of 12 glacial episodes occurred. These episodes extended to the close of the early Pleistocene (850 000 years ago). About 2 million years ago can, therefore, be taken as the beginning of the Ice Age.

The effects of the Ice Age deepened in the middle and late Pleistocene. In the interval between 850 000 years ago and 14 000 years ago there were at least eight large glacial episodes. The shift in emphasis from many small to a few large episodes is thought to have been triggered by the extension of permanent ice sheets to entirely cover the Arctic Ocean. Global cooling was thus accentuated and increasing amounts of ice piled up on the land. As a result, glacial periods were intensified and prolonged.

During a glacial episode, as snow and ice accumulates on the surface of the earth it reflects more and more of the sun's heat back into space. This heat is lost to the earth. Growing ice sheets push more pack ice and icebergs into the oceans and so cool the water. Changes of sea level accompanying ice accumulation affect the distribution of land areas and this in turn affects air and oceanic currents and snow and ice formation. Once a period of cold has been started it tends to be intensified and perpetuated by its own physical effects. But a turning point is eventually reached, possibly resulting from fluctuations in solar energy, atmospheric carbon dioxide, or tilting of the earth's axis, and a period of warming-up starts, with similar acceleration of physical effects. As melting begins to outpace the addition of new snow, the spread of ice stops. The ice sheets gradually thin, and the ice slowly begins to withdraw. As the climate becomes warmer, plants and animals migrate into areas abandoned by the shrinking ice sheet. The hardiest, or those able to spread most rapidly, come first. Meanwhile the ice dwindles and vanishes, or perhaps only a few mountain glaciers and isolated ice caps are left. We are now in an interglacial.

The climate of the interglacial periods was as warm as that of the present day, or warmer; and many warmth-loving plants and animals are recorded in interglacial beds far beyond their present-day range. For example, the hippopotamus was able to migrate into the British Isles, and in New Zealand kauri forests and tropical–

subtropical soils (coloured bright red) extended south to Wellington. But eventually, with the continuing swing of the pendulum, climate deteriorates and once more ice begins to accumulate. In this way, Pleistocene climates oscillated between cold glacials and warm interglacials.

Although the extent of ice accumulation varied considerably from one glacial period to another, at a period of maximum extent about 20 000 years ago, ice covered some 27% of the earth's land surface, compared with about 10% today. In the Northern Hemisphere extensive ice sheets formed on the land. An ice sheet centred on Scandinavia pushed southwards across northern Europe. Ice sheets covered the Urals and Siberia. The European mountain chains had their own ice sheets. In Britain, ice pushed down from Scotland, the Lake District, Wales, and central Ireland to cover all the British Isles except that part south of a line between London and Bristol. Both the Irish Sea and North Sea were filled with ice. In North America ice covered all of Canada and pushed southwards into the middle states of U.S.A. South of the main North American ice sheet ice caps developed on the higher mountain ranges. Elsewhere in the world ice formed on the mountains of South America, Africa, Australia, and New Zealand. Extensive areas of floating pack ice and icebergs blocked the Arctic and North Atlantic seas and pushed out on all sides from the massive ice caps on Antarctica.

The difference in mean annual temperature between glacial and interglacial episodes

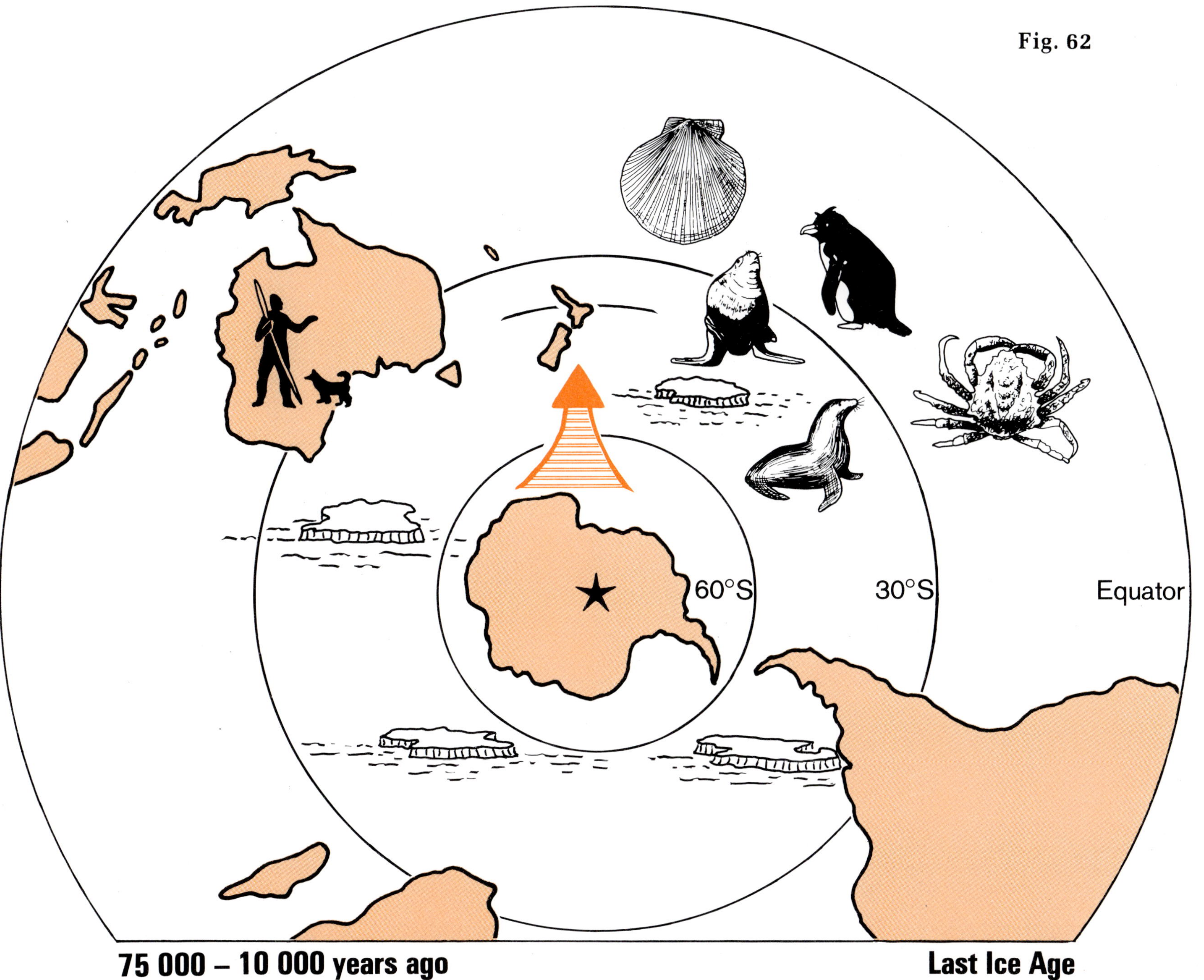

Fig. 62
60°S
30°S
Equator
75 000 – 10 000 years ago
Last Ice Age

was about 14°C. Where a region is already all ice, further cooling has little obvious effect and similarly, where the climate is already tropical, a small drop in temperature may affect vegetation and perhaps rainfall, but little else. But these temperature drops caused dramatic changes in middle latitudes. Mid latitude countries like New Zealand experienced a considerable swing in climate; many parts of the country passed from temperate conditions to those more like that of the treeless barren tundra regions of today (Fig. 63). However, the moderating influence of the sea meant the overall effects of glacial climates were not quite so severe in New Zealand, compared with continental Europe and North America. Auckland's climate probably changed to match that of Dunedin today, and Wellington experienced a frost climate similar to that of northern Norway today. In the North Island, the higher mountains had ice caps and small glaciers filled the upper reaches of some valleys in the Tararua and Ruahine ranges. Ice covered the higher ranges of the South Island, and extensive glaciers formed in the Southern Alps. The glaciers reached the sea along the West Coast and spread along the inland flanks of the Canterbury Plains. A thick ice sheet developed in Fiordland and pushed out westwards to the sea and eastwards towards Central Otago (Fig. 63).

Parts of New Zealand south of a line passing through the Waikato and the Bay of Plenty had a cold climate, with varying degrees of frost. In these areas it was not quite cold enough, nor possibly wet enough, for actual ice sheets to accumulate. The cold frosty climate was sufficiently severe to discourage the growth of forest, however, and many areas were either completely bare, or thinly covered with alpine-type vegetation and hardy grasses, rushes, and sedges. Forest

was restricted to coastal "refuge" areas and to the northern part of the North Island (Fig. 63). Parts of the continental shelves, exposed by falling sea levels during glacial periods, also acted as refuge areas.

The accumulation of snow and ice on the land locked up vast amounts of water that otherwise would have found its way back to the sea. Each glacial period was, therefore, accompanied by lowering of world sea levels. As the ice caps increased in extent and thickness so sea level dropped. During the last glacial period (30 000–20 000 years ago), world sea level was 105–135 m lower than it is now. Between 17 000 and 14 000 years ago it dropped even further, to reach 183 m below modern sea level. During times of low sea level in the early part of the last glacial it became possible for ancestors of the Aborigines to move from Southeast Asia across Indonesia into Australia (Fig. 62). It was much later that New Zealand received its first humans. About 1000 years ago Polynesian migrants sailed in canoes across the Southwest Pacific Ocean to colonise the islands of New Zealand.

The drops in sea level drastically altered the shapes of many of the present landmasses. Britain, for example, was joined to France, Belgium, Holland, and Germany across what is now the North Sea. The Bering Strait was dry and formed a land bridge for animal migrations between the Old (Asia) and New World (America). The myriad islands in Indonesia were all connected by land and linked to the mainland of Southeast Asia. Around the New Zealand coast, the retreating sea exposed broad areas of sea floor. The North and South Islands were joined across Cook Strait, and Stewart Island was linked across Foveaux Strait to Southland (Fig. 63).

Fig. 63 *(Left)* New Zealand in the last glacial phase of the Ice Age (the late Otiran glacial), 20 000–18 000 years ago. The ice-age shoreline is shown by the thick line, and the modern shoreline by the thin line. Snowfields and other areas above the snowline and beneath glacier ice are shown by the dots. The distribution of river gravels built up as a result of ice-age erosion is shown by the diagonal pattern. Volcanoes active at about this time are shown by the star symbols.

(Right) The probable distribution of vegetation in New Zealand in the last glacial phase of the Ice Age (the late Otiran glacial). Many of our forest plants (rimu, totara, etc.) were confined to the North Island north of Auckland, although small patches may have existed along some parts of the shoreline, where conditions were slightly warmer as a result of the moderating influence of the sea. Beech forest (i.e., cool-temperate forest) was very widespread and probably also covered large areas of the present sea bed which was then exposed by low sea levels.

AFTER THE ICE AGE

Retreat of the ice in New Zealand is thought to have started about 14 000 years ago. In many parts of the world ice sheets lingered until 10 000 years ago, so this latter date is often taken as the "official" end of the last glacial. It also marks the end of the Pleistocene period, or Ice Age, and the beginning of the Holocene. As the climate warmed and ice sheets melted, the sea rose from its low glacial levels (135–183 m below present) to reach about 75 m below present sea level 10 000 years ago. Between 10 000 and 5500 years ago the sea rose at a rate of about 1 m per century, until about 5500 years ago it had reached about the present level. The last remnants of the Scandinavian ice sheet had melted by 7000 years ago, and those of the North American ice sheet by 5500 years ago. Sea level then continued to rise, until between 5500 and 4000 years ago, at the height of a warm period (Climatic Optimum), it reached a height 2–3 m above present sea level. Since then the sea has dropped to its present level.

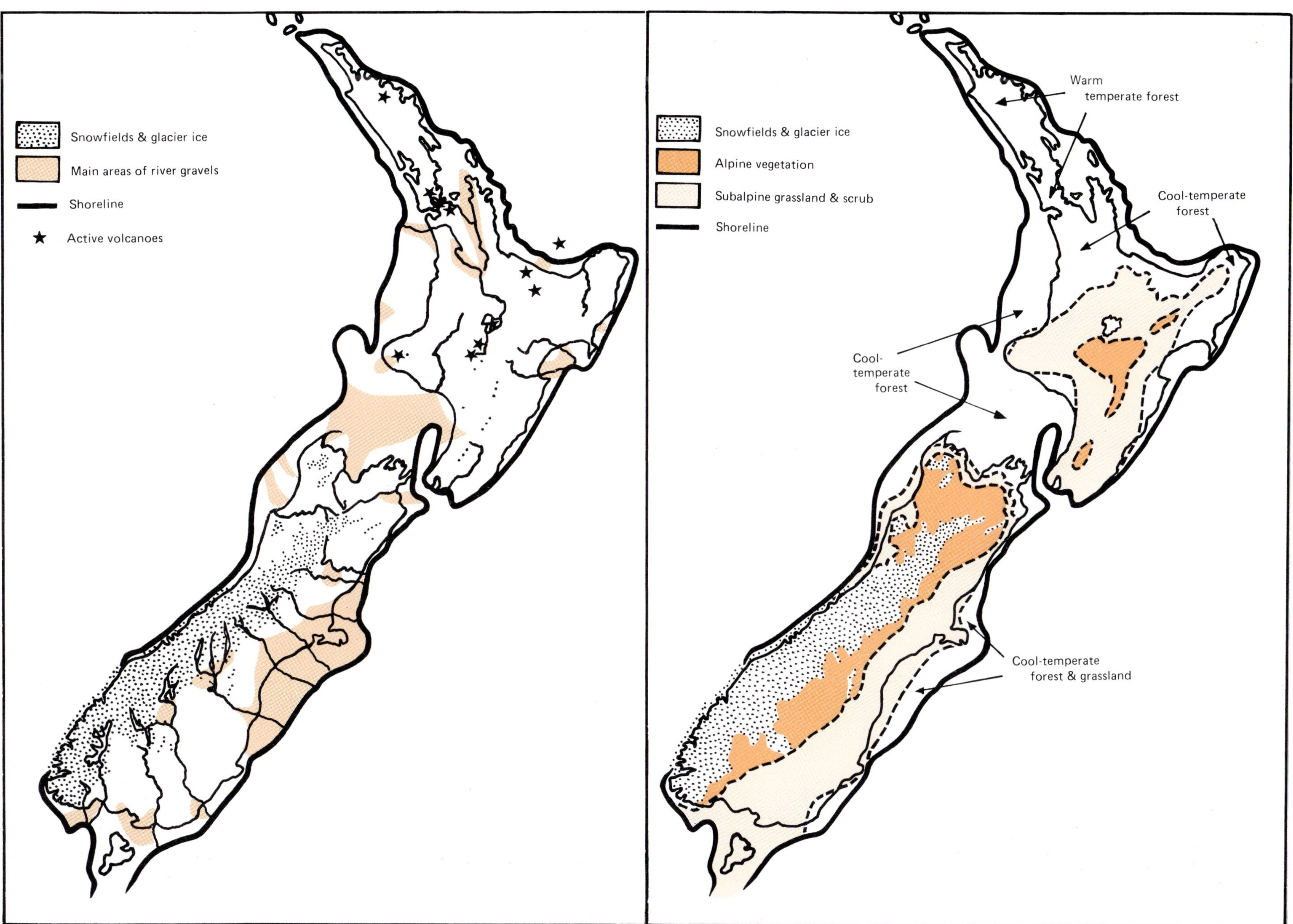
Snowfields & glacier ice
Main areas of river gravels
Shoreline
Active volcanoes
Snowfields & glacier ice
Alpine vegetation
Subalpine grassland & scrub
Shoreline
Warm temperate forest
Cool-temperate forest
Cool-temperate forest
Cool-temperate forest & grassland

PLEISTOCENE LIFE

The climatic fluctuations between cold and warm, as well as the emergence of modern man, had enormous effects on the flora and fauna of the Pleistocene. The rapid swings in climate forced plants and animals to migrate to remain in a suitable environment. Sometimes these migrations were readily accomplished, especially in those regions where the main barriers to migration (mountains and coastlines) extended in a north–south direction. In other regions the forced migrations led to extinctions, especially where organisms were forced up against a barrier, without avenues of escape.

Because of the relatively short duration of the Pleistocene and Holocene, plants show little evolutionary change and the changes seen in floras generally relate to climatic and human factors. The vertebrate record, in strong contrast to that of plants, shows very striking evolutionary changes. These are probably the result of intense selection pressures during this period of rapidly changing environments. In as little as 5000 years, areas adjacent to ice sheets could change, as the ice withdrew at the close of a glacial episode, from treeless tundra to temperate forest. We see evidence of rapid evolution, especially in the large vertebrates which adapted to a great range of ecological niches. Among the elephants, for example, the mammoth evolved, highly adapted to life in the cold. But it disappeared about 12 000 years ago, as climate warmed and as hunting pressures from man increased.

The beginning of the Pleistocene is marked by development of three important groups of mammals: modern elephants, modern one-toed horses, and modern bovines (now represented by cattle, bison, and buffaloes). Elephants and bovines initially evolved in the Old World. Horses, however, are of American origin and in the late Pliocene and earliest Pleistocene migrated into Eurasia via a land bridge across Bering Strait.

This land bridge was probably available during each glacial episode, when global sea level dropped and exposed vast areas of the sea floor. Each opening of the land bridge facilitated the movement of animals between the New and Old Worlds. Such movements were restricted to animals that were able to withstand the cold conditions on the land bridge. These ranged from extreme cold (tundra and glaciated areas) to probably cool-temperate at best, with pine and spruce forests. The mammoths, for instance, but not the elephants, were able to cross. Horses crossed, but rhinoceroses did not. Animals such as camels (which, contrary to popular opinion, are quite hardy northern animals), reindeer, moose, wapiti, bison, mountain goat, and sheep made the journey, while such forms as giraffes, deer, and most antelopes and cattle were unable to cross. Some immigrants from the Old World scarcely spread beyond the North American end of the land bridge (e.g., the musk ox, the yak, and the saiga antelope).

Two groups of elephants evolved in the northern continents; the mammoths and the forest elephants. The mammoths lived on tundra vegetation: steppe grasses and herbs as well as twigs and leaves of birch and pine. The woolly mammoth had a unique appearance. It was immense, with strongly curved tusks (which were used to scrape the snow off the grass), a peaked head, a humped, sloping back, and a thick woolly coat of hair. Some reached a shoulder height of up to 4.5 m. The straight-tusked forest elephants were up to 4 m in shoulder height, and the tusks were nearly straight and of enormous size. This type of elephant inhabited the open forests and plains of Eurasia and was a characteristic member of the interglacial fauna in Europe. Foreign to the tundra, it was unable to cross the Bering Strait land bridge and so remained confined to the Old World.

Before the Pleistocene, the ancestors of modern humans had evolved along several lines. After separation in Miocene times from lineages that eventually gave rise to the apes (chimpanzee, gorilla, orangutan, and gibbon) metre-tall ape-like *Australopithecus* appeared in the Pliocene (3.0–3.5 million years ago). Somewhat taller tool-making humans, *Homo habilis*, appeared in the earliest Pleistocene about 2 million years ago. Modern man, *Homo sapiens*, appeared in the late Pleistocene 500 000–250 000 years ago. One of the most numerous groups of late Pleistocene humans was Neanderthal man. His heavy eyebrow ridges and sloping forehead give him a rather ape-like appearance. Although his average height was only around 1.5 m he had a brain capacity greater than in modern man. However, the brain itself seems to have been less complex, indicating that he was probably not as intelligent.

Another early group of humans was Cro-Magnon man, physically indistinguishable from modern man, and responsible for the famous cave paintings of northern Spain and southern France.

Throughout the Pleistocene humans had hunted in small bands and occupied small seasonal camp sites. However, about 10 000 years ago, as climate improved and the Pleistocene ended, humans living around the Mediterranean and in the Middle East began to occupy permanently established villages. At about this time the dog was domesticated and presumably helped with hunting. In the early Holocene (9000–7500

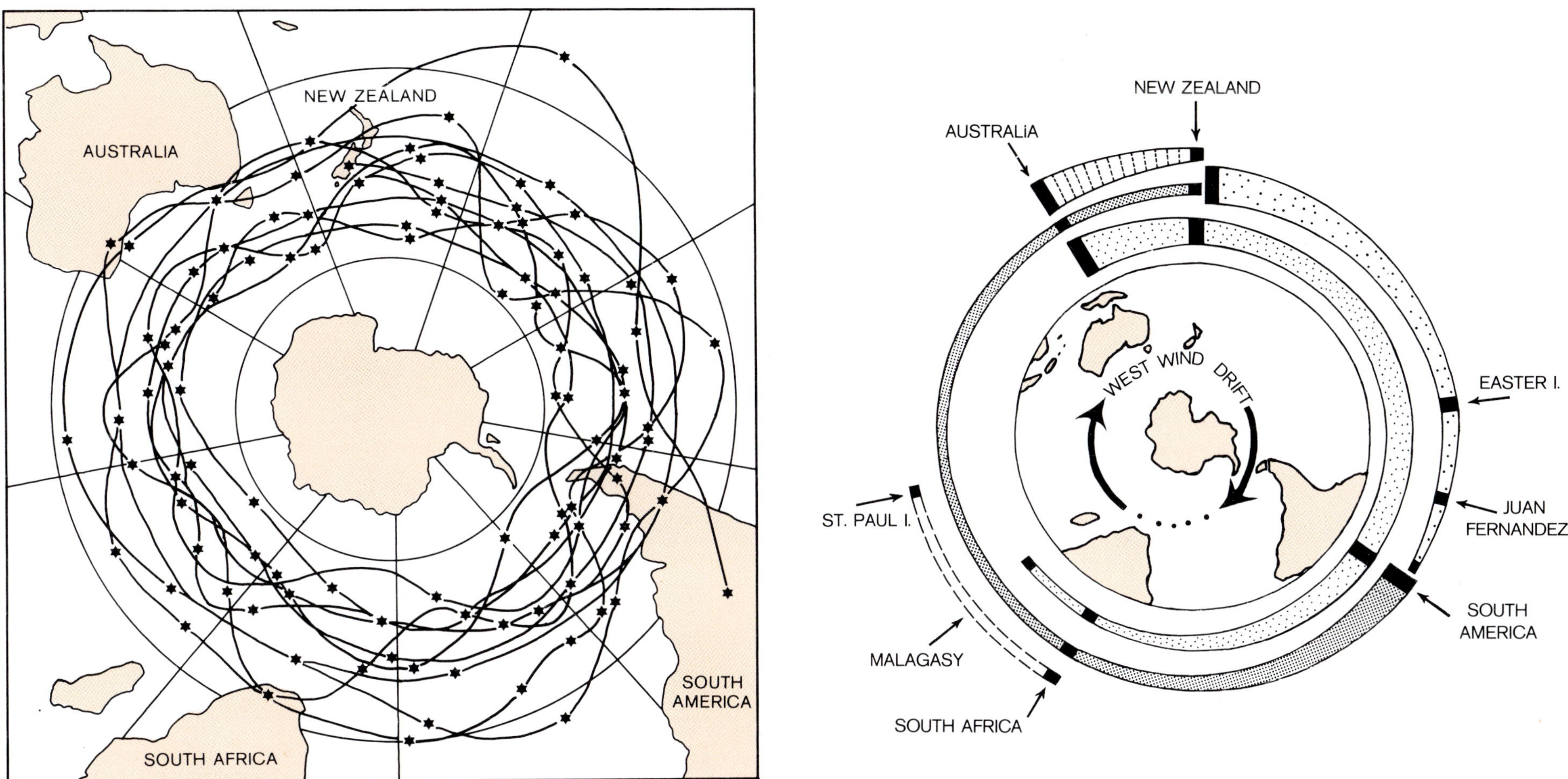

years ago), in the eastern Mediterranean, varieties of wheat and barley were regularly sown and harvested and sheep and goats were kept in enclosed pastures. In the late Holocene (6000–3000 years ago), trading centres (towns) became established in the eastern Mediterranean, a wide range of cereals was grown, and hunting had largely given way to the care of domestic animals (sheep, goats, cattle, pigs, and donkeys).

Fig. 64   The prevailing westerly winds of mid latitudes in the Southern Hemisphere (the "West Wind drift") have had, and are still having, an enormous influence on the flora and fauna of southern lands (including New Zealand). The two diagrams above provide graphic illustrations of the effects of the West Wind Drift. The diagram (*left*) depicts the track of a large weather balloon, released at Christchurch and tracked by satellite for 102 days. After release, the balloon reached South America in just over 5 days. In the 102 days it was tracked, it circled the world 8 times and repeatedly crossed all of the southern continents and most islands. The stars indicate days elapsed from time of launch. The balloon tracks serve to emphasize the constancy and strength of the West Wind system. The diagram (*right*) illustrates how the West Wind Drift has influenced the distribution of echinoderms (sea stars, sea eggs, etc.) in southern latitudes. Each land in the path of the West Wind Drift has a "tail" of wind-drifted echinoderms, extending in the direction in which the West Wind is blowing. In each example the height of the dark vertical bars is proportional to the number of varieties (species) of a particular kind (genus) of echinoderm present in the indicated land areas. An example of each type of distribution is included: Australian–New Zealand; New Zealand–South America; Australia–New Zealand–South America–South Africa; South America–South Africa–Australia–New Zealand; South Africa–St. Paul Island.

PLEISTOCENE LIFE IN NEW ZEALAND

Steady deterioration of climate in late Miocene and Pliocene times thinned out many of the warmth-loving immigrants New Zealand had received in earlier ages. The *coup de grace* was delivered by the severe climates of the Pleistocene glacials, occurring between 2 million and 10 000 years ago. During a number of these glacials temperate organisms were restricted to northernmost New Zealand and to a few coastal refuges where the influence of the sea moderated the cold glacial climate. There was no escape northwards beyond latitude 35°S (beyond North Cape; Fig. 63), so decimation was frequently the lot of warmth-loving organisms, and many disappeared completely, never to return. Examples included plants such as *Casuarina*, *Acacia*, and the *brassi* group of *Nothofagus*.

The northward retreat of warm and temperate organisms was matched by northward advance of those with cold-temperate requirements. Thus *Chlamys delicatula*, a subantarctic scallop-like bivalve, now found south of Stewart Island (latitude 47°S), advanced north into Hawkes Bay (latitude 39°S) in the early Pleistocene. The spider crab *Jacquinotia*, now found around Campbell Island and Auckland Islands, moved a comparable distance northwards at the same time. Other cold water organisms such as sea lions, seals, and penguins showed similar northerly extensions of their latitudinal ranges. The drawings on Fig. 62 depict (clockwise from top) some of the cold water migrants that moved northwards in the sense indicated by the broad arrow: *Chlamys delicatula*, sea lion (*Neophoca*), penguin (Eudyptes), Ross seal (*Ommatophoca*), spider crab (*Jacquinotia*).

During the rigorous climate of the last glacial phase (65 000–10 000 years ago) podocarp forests largely survived in northern areas beyond Hamilton, and in a few refuges at sea level (Fig. 63). Survival of the podocarps, and the consequent continuation of supply of their fleshy seeds (very attractive to birds and other animals), in turn ensured the survival of many of the archaic land animals dating back to the time New Zealand was part of Gondwanaland.

The changing patterns of land and sea that were a feature of Pleistocene time encouraged development of distinctive varieties of plants and animals, as a result of isolation. Distinct geographic races developed in native land snails, for example, as populations became cut off by movements of land and sea. The Three Kings Islands, immediately north of New Zealand, remained completely isolated, even during the times of lowest glacial sea level and as a result now have a number of distinctive plants and animals.

As alpine regions came into being, New Zealand developed its own alpine and sub-alpine races from organisms of Malayo-Pacific (i.e., sub-tropical) origin. In the early Pleistocene, as many of the New Zealand mountains began to be pushed up, certain plants and animals moved up with them, especially those adapted to life in dry stony river beds, rocky slopes, etc. Thus the alpine rock wren, alpine wetas, and alpine cicadas were developed from lowland groups, and the New Zealand parrots produced the alpine kea. (The kaka is its equivalent in lowland forests.)

As climate warmed after the last glacial phase some 10 000 years ago, some of the gaps in the New Zealand flora and fauna (resulting from Pleistocene extinctions) were filled by temperate organisms riding the West Wind Drift (Fig. 64). Forest gradually became re-established throughout New Zealand, but its recovery from the repeated disruptions during the successive glaciations was a long and slow process. It is believed that even within the span of man's occupation of New Zealand, vegetation changes have occurred which are related to this long-term recovery process. Coupled with this, are the effects of climatic changes: notably the Climatic Optimum (a warm period extending from 7000–4000 years ago), and the Little Ice Age (a cold period between A.D. 1550–1880).

Without a doubt the arrival of Polynesian man about a thousand years ago initiated a long train of biological events, that continued even more rapidly after the visits of Tasman and Cook, and the arrival of European settlers. The early Polynesians (the "Moa Hunters") used the moa and other birds as easily-hunted protein sources, and so deprived the New Zealand fauna of many of its older distinctive elements (including the moa, and native New Zealand geese, swans, eagles, crows, etc. (Fig. 65). Fire brought by human beings, and used by them in hunting and agriculture, destroyed large areas of forest in coastal and central North Island and eastern South Island. This reactivated many areas of hitherto stable sandy country so that sand dunes invaded fertile land in many coastal regions.

The Polynesian rat and dog (also introduced) added to the effects of hunting and use of fire. European settlers introduced, by accident or design, a wide variety of animals and plants from other parts of the world which competed, often successfully, with native species. In particular New Zealand's formerly abundant bird life was decimated. Many birds disappeared altogether, while others became restricted to Fiordland and various islands off the New Zealand coast. Other animals such as tuatara, native frogs, wetas, and native land snails, apparently formerly widely distributed over both North and South Islands, became greatly reduced both in numbers and geographic distribution.

Thus the arrival of the human race, with fire, rats, and dogs, coming on top of the effects of the Ice Age, sounded the death-knell for many of New Zealand's unique primaeval organisms, some dating back tens of millions of years to times when New Zealand was part of the now fragmented great southern continent of Gondwanaland.

Fig. 65 The arrival of man about a thousand years ago sounded the death-knell for many elements of the New Zealand native fauna. The early Polynesians ("moa hunters") hunted many of the larger varieties of moa into extinction, along with native varieties of geese, swans, eagles, and crows. Evidence for the presence of these animals in moa-hunter times comes from camp sites and deposits laid down in swamps and caves. The Pyramid Valley swamp, north Canterbury, yielded a treasure trove of skeletons, including those of the giant moa, *Dinornis maximus*, standing 3 m tall. The swamp, some 1.5 hectares in area, holds the bones of an estimated 2800 moa. The large birds were presumably bogged after falling through the deceptively thin swamp crust. Remains of the New Zealand eagle, *Harpagornis*, have also been found in the swamp. These eagles were probably feeding on trapped moa, and became victims themselves. *Harpagornis* was a large and powerful predator, with a massive skull, a menacing hooked beak, and a wingspan of 3 m. It was probably the heaviest eagle known to science.

PLEISTOCENE ROCKS IN NEW ZEALAND

The Pleistocene record of New Zealand can be divided into two segments: an earlier, largely marine, phase and a later largely terrestrial phase. The earlier marine phase extended from about 2 million years ago to about 320 000 years ago. During this time, many of the areas that had been under the sea during the Pliocene (Fig. 60) remained submerged and were sites for deposition of a variety of marine sediments. In these areas Pliocene and early Pleistocene rocks were laid down as virtually continuous sequences, without major breaks.

Marine deposition took place in the Bay of Plenty region, Poverty Bay, Hawkes Bay, Wairarapa, Wanganui, Marlborough, and northern Canterbury. In other low-lying areas deposits of early Pleistocene age were built up as material was eroded from rising mountains. Volcanic activity in Waikato, Coromandel, Bay of Plenty, and Taranaki contributed material to the adjacent lowland areas. Substantial amounts of rock debris also came from those areas affected by glaciation and frost activity. Gravels, sands and silts from the South Island alpine regions filled low-lying areas in Nelson and Marlborough, along the entire length of the West Coast and Canterbury Plains, and in Southland.

The seas that covered parts of New Zealand in the early Pleistocene were generally not deep and the gravels, sands, and muds laid down contain abundant shells. They were often deposited either immediately offshore or in depths of 20–50 m along the shallow inner margin of the continental shelf. In some areas, however, water depths were greater, extending to the deeper outer edge of the continental shelf (100–200 m) or down the continental slope into bathyal depths. Thick "papa" deposits of siltstone and mudstone were often laid down in such deep water areas.

Active uplift of New Zealand continued into the Pleistocene, and continues today (Fig. 66). Many mountain ranges of today's landscape were already largely formed by the end of the Pliocene, particularly those of the South Island.

The Kaweka, Ruahine, Tararua, and Rimutaka Ranges were slower to rise out of the sea, and most of their growth occurred in the last million or half million years.

The seas covering Taranaki–Manawatu and Hawkes Bay–Poverty Bay areas in the Pliocene and early Pleistocene were linked by two seaways extending across the site now occupied by the Kaweka, Ruahine, and Tararua ranges. A variety of marine sediments were laid down in these seaways. The northern seaway, over the Kuripapango area of the Napier–Taihape highway, remained open until some time in the middle part of the Pleistocene (about 1 million years ago), although it was progressively narrowed by rising land on either side. The southern seaway, over the area now occupied by the Manawatu Gorge, also remained open, although it too was gradually thinned down to form a narrow strait. Then, as uplift continued and accelerated (between 1 million and 630 000 years ago), both straits became choked with coarse gravel and sand fed from either side, and the sea retreated. After the sea had been excluded, a large river system continued to follow the course of the old Manawatu seaway. As uplift of the main ranges continued, the rivers coalesced into one, and this major river, greatly augmented in volume, was able to maintain its course across the rising mountains by cutting down and forming the Manawatu Gorge we see today.

Although both the Hawkes Bay and Manawatu seaways lost their battle with the rising land, the sea has managed to keep Cook Strait open. During times of low glacial sea level, however, especially during the last glacial period, large areas of western Cook Strait were dry land (Fig. 63). A continuous plain stretched between Nelson and the Taranaki–Wanganui area, cutting off Cook Strait from the Tasman Sea. Old beach deposits, laid down along the coastline of the time, have been charted offshore between Nelson and Taranaki, and 18 000-year-old moa bones have been found in the strait, indicating that moas roamed across the now submerged plain.

As mountain building continued the sea was gradually pushed off the rising land and from about 320 000 years ago onwards, the New Zealand late Pleistocene and Holocene record is mainly preserved in the form of terrestrial deposits built up by glacial, frost, and alluvial processes. Glacial deposits flank the Southern Alps on all sides, forming festoons of moraines, snaking back and forth across the landscape. Elsewhere, marginal to the Southern Alps and in the southern half of the North Island, bare rock surfaces were shattered and crumbled by intense frost activity and the countryside often became thickly mantled with deposits of rubble. Rock glaciers composed of such material flowed down valley sides and into low-lying depressions.

Fig. 66 These sketches give an idea of the general nature and sequence of events that affected the New Zealand region between the early Permian (300 million years ago) and the present day. They are very generalised and do not attempt to show the many very complex situations that developed during crustal movements in the southwest Pacific (compare Fig. 20).

A and B show the situation in early Permian–late Jurassic times. The area now occupied by New Zealand was part of the sea floor, receiving sediments (gravel, sand, mud) from the eroding margins of eastern Australia and Antarctica. Volcanoes on both land and sea floor contributed lava and volcanic ash. In diagram C, earth movements heralding formation of the Tasman Sea and Southern Ocean have compressed the New Zealand region. The accumulated sedimentary material was squeezed and folded upwards to form an ancestral New Zealand landmass, extending northwards to New Caledonia and southwards to Campbell and Auckland Islands. Rivers, streams, and the sea began to erode the new landmass, and the products of erosion were deposited in the sea to either side (D).

By middle and late Cretaceous time (100–65 million years ago), some parts of ancestral New Zealand were worn down almost to sea level (diagram E), and the sea was beginning to invade the land. Erosion of ancestral New Zealand continued until most of the topographic relief had been worn away. In the early

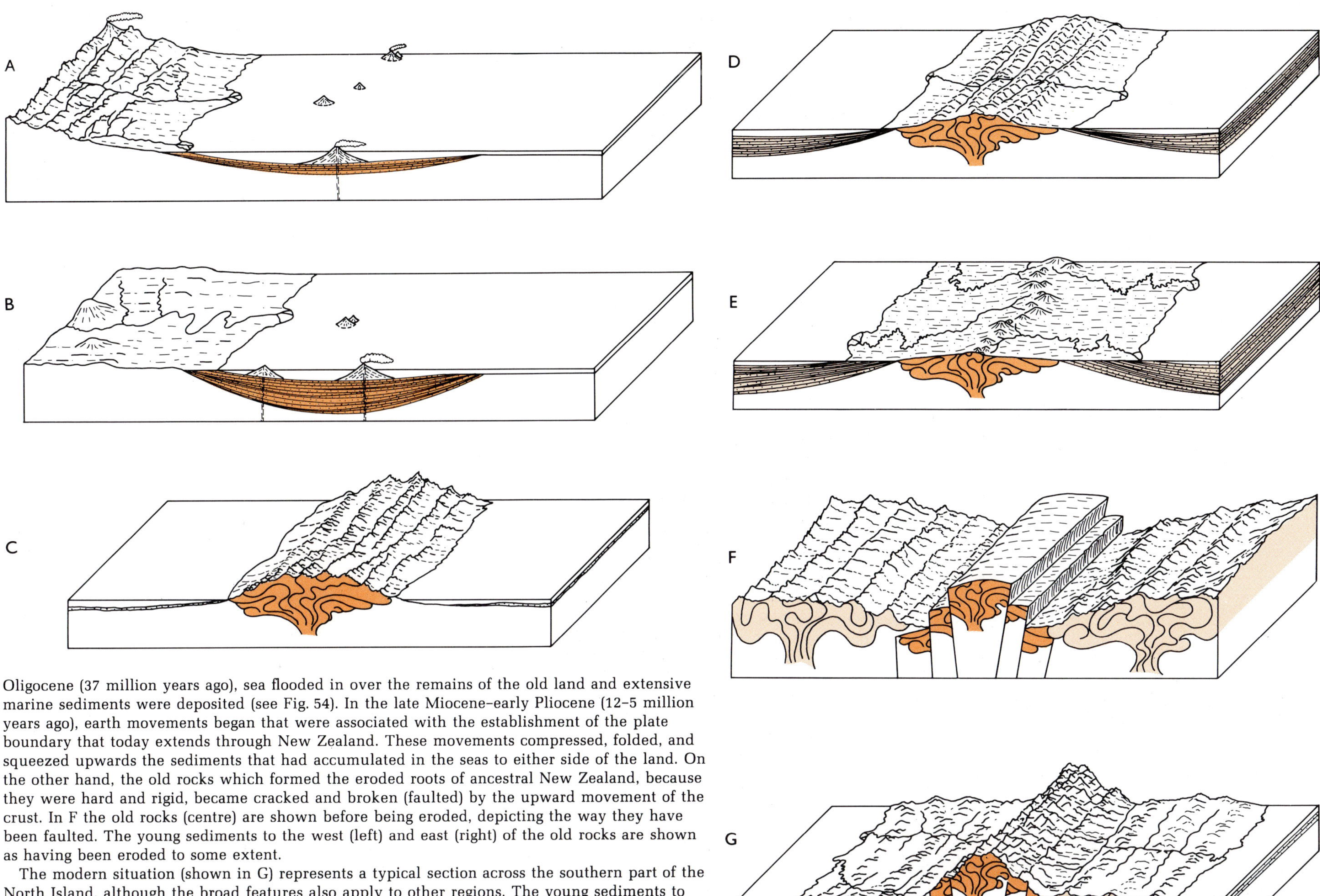

Oligocene (37 million years ago), sea flooded in over the remains of the old land and extensive marine sediments were deposited (see Fig. 54). In the late Miocene–early Pliocene (12–5 million years ago), earth movements began that were associated with the establishment of the plate boundary that today extends through New Zealand. These movements compressed, folded, and squeezed upwards the sediments that had accumulated in the seas to either side of the land. On the other hand, the old rocks which formed the eroded roots of ancestral New Zealand, because they were hard and rigid, became cracked and broken (faulted) by the upward movement of the crust. In F the old rocks (centre) are shown before being eroded, depicting the way they have been faulted. The young sediments to the west (left) and east (right) of the old rocks are shown as having been eroded to some extent.

The modern situation (shown in G) represents a typical section across the southern part of the North Island, although the broad features also apply to other regions. The young sediments to either side were soft and relatively unconsolidated, and were rapidly eroded to form lowlands such as Manawatu-Wanganui (left) and Hawkes Bay (right) regions. The old, hard rocks, because they have been pushed up along faults, stand up as the main ranges of today.

Sources of illustrations

The continental reconstructions depicted in the paleogeographic maps have been based on:

Firstbrook, P. L.; Funnell, B. M.; Hurley, A. M.; Smith, A. G. 1979. Paleoceanic reconstructions 160–0 Ma. National Science Foundation.

Scotese, C. R.; Bamach, R. K.; Barton, C.; van der Voo, R.; Ziegler, A. M. 1979. Palaeozoic base maps. *Journal of Geology 87* : 217–277.

Smith, A. G.; Briden, J. C. 1977. Mesozoic and Cenozoic paleocontinental maps. Cambridge University Press.

The individual paleogeographic maps of New Zealand are simplified versions of maps prepared for "The Geology of New Zealand" (Editors: R. P. Suggate, G. R. Stevens, M. T. Te Punga), published in 1978 by New Zealand Geological Survey.

The reconstructions of animal life in Fig. 39 and 45 are based on paintings originally published in "Prehistoric Animals" by J. Augusta and Z. Burian (Spring Books, London 1958) and "Prehistoric Sea Monsters" by J. Augusta and Z. Burian (Spring Books, London 1964); Fig. 61 on a painting published in "The Age of Birds" by A. Feduccia (Harvard University Press, 1980); Fig. 65 on a painting published in "Pyramid Valley" by R. Duff (Canterbury Museum, 1949).

Further reading

The following references are sources of general background material:

Blackett, P. M. S. et al. 1965: A symposium on continental drift. Royal Society, London.

Calder, N. 1972: Restless earth. British Broadcasting Corporation, London.

Gass, I. G.; Smith, P. J.; Wilson, R. C. L. *ed.* 1971: Understanding the Earth. Artemis Press, Chichester.

Hallam, A. 1973: A revolution in the earth sciences: From continental drift to plate tectonics. Oxford University Press.

Institute of Geological Sciences 1972: The story of the earth. Her Majesty's Stationery Office, London.

Laporte, L. F. et al. 1978: Evolution and the fossil record. Scientific American Books (W. H. Freeman, San Francisco).

Marvin, U. B. 1973: Continental drift: The evolution of a concept. Smithsonian Institution Press, Washington D.C.

Matthews, S. W. 1973: This changing earth. *National geographic magazine 143(1)* : 1–37.

Menard, H. W. et al. 1977: Ocean science. Scientific American Books (W. H. Freeman, San Francisco).

Moore, J. R. *ed.* 1972: Oceanography. Scientific American Books (W. H. Freeman, San Francisco).

Oxburgh, E. R. 1974: The plain man's guide to plate tectonics. *Proceedings Geologists' Association 85* : 299–357.

Phinney, R. A. *ed.* 1968: The history of the earth's crust: A symposium. Princeton University Press.

Press, F.; Siever, R. 1974: Earth. W. H. Freeman, San Francisco.

Revelle, R. et al. 1969: The ocean. Scientific American Books (W. H. Freeman, San Francisco).

Runcorn, S. K. *ed.* 1962: Continental drift. Academic Press, London.

Sullivan, W. 1974: Continents in motion: The new earth debate. Macmillan, London.

Tarling, D. H.; Tarling, M. P., 1971: Continental drift. Bell, London.

——— 1973: Continental drift: a study of the Earth's changing surface. Pelican, London.

Tarling, D. H.; Runcorn, S. K. *ed.* 1973: Continental drift, sea floor spreading and plate tectonics: Implications to the earth sciences. (2 volumes). Academic Press, London.

Wilson, J. T. et al. 1972: Continents adrift. Scientific American Books (W. H. Freeman, San Francisco).

Wilson, J. T. *ed.* 1976: Continents adrift and continents aground. Scientific American Books (W. H. Freeman, San Francisco). Wyllie, P. J. 1971: The dynamic earth. Wiley, New York.

——— 1974: Plate tectonics, sea-floor spreading and continental drift: an introduction. *American Association of Petroleum Geologists memoir 23* : 5–15.

——— 1976: The way the earth works: An introduction to the new global geology and its revolutionary development. Wiley, New York.

Detailed treatment of New Zealand topics is provided in:

Fleming, C. A. 1962: New Zealand biogeography: a palaeontologist's approach. *Tuatara 10* : 53–108.

——— 1975: The geological history of New Zealand and its biota. Pp. 1–86 *in* Kuschel, G. *ed.* Biogeography and ecology in New Zealand. W. Junk, The Hague.

——— 1977: The history of life in New Zealand forests. *New Zealand journal of forestry 22*: 249–262.

——— 1979: The geological history of New Zealand and its life. Auckland University Press.

Hatherton, T. 1978: Astride two plates: an account of current tectonics in New Zealand. *Endeavour 2*: 180–185.

Molnar, R. E. 1981: A dinosaur from New Zealand. Pp. 91–96 *in* Cresswell, M. M.; Vella, P. *ed.* Gondwana Five. Rotterdam, A. A. Balkema.

Stevens, G. R. 1977: Mesozoic biogeography of the Southwest Pacific and its relationship to plate tectonics. Pp. 309–326 *in* International symposium on geodynamics in the Southwest Pacific. Editions Technip, Paris.

——— 1980: New Zealand adrift: The theory of continental drift in a New Zealand setting. A. H. & A. W. Reed, Wellington.

——— 1980: Southwest Pacific faunal biogeography in Mesozoic and Cenozoic times: a review. *Palaeogeography, Palaeoclimatology, Palaeoecology 31* : 153–196. (And other papers in this issue.)

Walcott, R. I. 1978: Present tectonics and late Cenozoic evolution of New Zealand. *Geophysical journal, Royal Astronomical Society 52* : 137–164.

Acknowledgments

This publication is based in part on the 48th Thomas Cawthron Memorial Lecture ("Our Wandering Islands: A New Zealand View of the Theory of Continental Drift") presented under the auspices of the Cawthron Institute, Nelson. Figures 1–26, and the individual hemisphere maps, are the work of my wife Diane. Colour separations have been prepared by Stephan Spiekerman and Chris Kitto (Science Mapping Unit, DSIR). Figures 28, 34, 37, 39, 41, 45, 61, 65, and the drawings of animals and plants incorporated in the individual maps are the work of Ron Brazier (Paleontological Artist, New Zealand Geological Survey). Fay Tonks and Patricia White (New Zealand Geological Survey), transposed the manuscript to word-processor. Ian Mackenzie and Quentin Ruscoe (Science Information Publishing Centre, DSIR) provided much appreciated assistance and encouragement in all aspects of publication.

Technical notes

The material presented in this book has been developed and considerably expanded from geological and paleontological themes published in "New Zealand Adrift" (A. H. & A. W. Reed 1980) and the 48th Thomas Cawthron Memorial Lecture (Cawthron Institute, Nelson, 1983). Like these two previous works, the present text and illustrations have been prepared for a non-technical audience and for this reason jargon and scientific theorising have been kept to a minimum. Throughout I have attempted to convey something of the stimulation and excitement of the bold new ideas in the earth sciences that have transformed the understanding of our planet.

In writing this book I have tried to present a coherent and readable account of the interwoven sequences of events that have shaped the modern world. However, in providing such an account I have been very aware of the many gaps in the geological record, and the fact that often these gaps must be bridged by speculation and extrapolation. Therefore widely differing interpretations may often be proposed in the scientific literature for the same set of observations. In dealing with the various interpretations expressed in the literature I have had to make fairly arbitrary decisions. However, a bibliography of key references has been included (p. 112) and these may be consulted for detailed discussion of the many alternative interpretations that for the sake of readability I have not been able to discuss in any great depth.

The text is based on scientific information and concepts current in the early 1980's. The nature of scientific progress means that new ideas, and new variations on old ideas, are being continually put forward in the scientific literature. However, just because ideas are new it does not necessarily mean that they are proved to be correct in the long run. They have to go through an often prolonged process of testing and evaluation, that they may, or may not survive. During the production phase of this book a number of new ideas have appeared in the scientific literature that if widely accepted have the potential of modifying some aspects of the material presented. It is, however, too early to tell whether such ideas will stand the test of time. Nonetheless, to alert readers to potential alternatives, reference is made to some of this new work below.

Continental reconstructions
(Fig. 24–26)

Since the advent of theories of plate tectonics, and before that of continental drift, a great variety of continental reconstructions have been proposed, each one based on different sources, and sometimes using different plotting techniques.

The continental reconstructions used in this book have been based on the compilations published by Smith & Briden (1977), Firstbrook et al. (1979), Scotese et al. (1979), and Smith et al. (1981). Other reconstructions have however been recently published, some of which attempt to reconcile the information obtained from various sources, notably from paleomagnetic studies of continental rocks and from models of evolution of the oceanic floors obtained from analyses of marine magnetic anomalies. These new reconstructions include those published by Barron et al. (1981), Irving (1983), Parrish et al. (1982), and Smith (1981, 1982).

The reconstructions used in this book and those cited in the references above are all based on the assumption that the diameter of the earth has remained constant throughout geological time. However, there are scientists who maintain that this assumption is not valid. They consider that the earth has been expanding at a constant rate throughout geological history. In the Jurassic, for example, the earth is proposed to have been only 80% of its present diameter. If this view is correct (and it is early days yet) the geographies of the world in the past are likely to be substantially different in detail from those illustrated in this book, although many of the general relationships would remain essentially unchanged.

Recently published continental reconstructions based on the theory of an expanding earth include those published by Carey (1976), King (1983), and Owen (1976, 1981, 1983a, b, c).

With a substantially smaller globe, the landmasses of the world may have been in much closer contact than shown on the maps in this book, allowing widespread interchange of animals and plants. This particularly applies to countries around the Pacific (e.g., Shields 1983), and bordering the ancient Tethys Ocean (e.g., Owen 1983c).

Another factor adding to the complexity of the task of reconstructing past continents is the possibility that some of the modern landmasses are composed of a patchwork quilt of individual fragments that have been rafted from various directions by seafloor spreading movements. According to proponents of this concept, displaced fragments, or "exotic terranes", have been identified in many areas, especially around the margins of the Pacific, and including New Zealand (Nur & Ben-Avaraham 1977; Howell 1980; Howell et al. 1984; Kamp 1980; Tozer 1982, 1984). It is, however, a subject still surrounded by considerable controversy (e.g., Retallack 1984).

Reconstruction of the New Zealand region (Fig. 21)

The reconstruction presented as Fig. 21 is based on the work of Grindley & Davey (1982). Later publications have maintained the general features of this reconstruction, but some variations in detail have been presented (Lock 1983; Grindley & Oliver 1983). New studies of stages in the evolution of the plate boundary extending through New Zealand (Fig. 22, 23) have been published by Walcott (1984) and Kamp (1984).

Geological time scale
(refer to endpapers)

Although the divisions and subdivisions of the geological time scale are generally accepted, the assignment of ages to their boundaries is evolving, with new refinements being introduced from a wide variety of sources.

The time scale adopted in this book is that used by Cohee, Glaessner & Hedberg (1978) and Stevens (1980). Since that time, however, a number of other time scales have been published, using differing calibration schemes (Harland et al. 1982; Odin 1982a, b; Odin et al. 1983; Palmer 1983). Therefore, in researching the geological literature the reader will undoubtedly find some variation in the dates assigned to the various geological boundaries of the geological scale, depending on the source used.

Ancestry of ratite birds (p. 58)

The origin of the unusual group of flightless birds the ratites (including the moa, kiwi, ostrich, emu, rhea, cassowary, and elephant bird) has been a contentious issue of long-standing in zoology and paleontology. The rarity of fossil ratites has meant that emphasis has had to be placed on interpretations based on zoological studies such as embryology and comparative morphology.

Two main possibilities have been proposed:
1. The ratites are a very ancient group of birds and have always been flightless. They evolved from reptiles before evolution of the flying birds (carinates).
2. The ratites did not evolve from reptiles independently and separately from all other birds, but rather have been derived from flying birds (carinates), and subsequently became specialised for a flightless way of life.

Although discussions about ratite origins have waxed and waned, new perspectives have recently been obtained from studies in embryology, comparative anatomy, and biochemistry. These studies have been reviewed by McGowan (1982, 1984) and Rich & Balouet (1984). On the balance of the evidence, McGowan considers that although the ratites have been derived from more primitive ancestral stocks than the flying birds, and are at least in this sense probably closer than flying birds to the reptiles (notably the theropod dinosaurs), they are not so primitive that they never passed through a flying stage. McGowan concludes that the ratites descended from the ancestral stock of birds sometime after the power of flight had been acquired, but before the modern flying birds developed.

Monotremes (pp. 52, 64)

Because the monotremes have left virtually no fossil record, at least for the critical early stages of their evolutionary development (e.g., Rich 1982), discussion of monotreme origins and migratory routes is entirely speculative. Nonetheless, the monotremes (and the marsupials) are of interest in the New Zealand context because of their apparent absence from New Zealand. This absence contrasts with the fact that both New Zealand and Australia shared in the radiation of ratite birds across the Gondwana continents that probably took place at a comparable time, presumably in the Jurassic and Cretaceous. The question thus arises: if ratites were able to reach New Zealand and Australia, why not monotremes and marsupials?

The viewpoint adopted in this book is that these differences between the land faunas of Australia and New Zealand are the result of differences in times of radiation of creatures such as ratites, monotremes, and marsupials, in relation to breaking-up movements leading to formation of the Tasman Sea, the Pacific and Southern Oceans.

A major factor in any discussion of monotreme origins is the divergence of zoological opinion on whether monotremes are mammal-like reptiles or reptile-like mammals. If viewed as reptilian they may be considered to be the only living survivors of the world-wide radiation of mammal-like reptiles that occurred in Triassic times (see reviews in Lillegraven 1979; Kemp 1982, 1983). If, on the other hand, monotremes are viewed as mammals they may have evolved much later, perhaps in early Cretaceous times (Kemp 1982, 1983). Because paleogeographic patterns, and opportunities for land migration, varied markedly between Triassic and Cretaceous times, the timing of radiation of the monotremes is of crucial importance. In the Triassic, the availability of land routes enabled a number of land animals to radiate widely across both Laurasia and Gondwana. By Cretaceous times, however, many of the land routes had been disrupted by areas of newly formed ocean.

Although both views of monotreme classification are presented in appropriate contexts in this book (pp. 52, 64), some of recent publications favour classification of the monotremes as mammals (Archer 1984a; Griffiths 1983; Strahan 1983). If this view gains wide currency the mono-tremes, because they are so different from other living mammal groups, can probably be considered to have diverged at a very early stage from the ancestral mammalian stock. In this context, it has been suggested that they were part of the radiation of the Multituberculata, a late Mesozoic–early Cenozoic group of mammals with large blade-like premolar teeth (Clemens 1979a; Murray 1984). In this scenario, as illustrated in Fig. 42, radiation of ancestral monotremes into the Southern Hemisphere probably took place in early Cretaceous, using the land connections that had been developed between many of the Gondwana continents at that time. The warm-temperate trans-Tasman route into New Zealand had been disrupted by this time, however, and the only available route (used by proteas and *Nothofagus*) was a southern route, probably with cool- or cold-temperate characteristics. Climatic conditions along such a route and links with South America/West Antarctica (rather than Africa and Eurasia) may have been factors preventing its use by ancestral monotremes. As a consequence, neither ancestral monotremes (nor marsupials) were able to enter New Zealand.

An alternative hypothesis is that monotremes at least, if not marsupials, were originally in New Zealand, but were eliminated during times of reduced habitat (e.g., the marked reduction of land area that occurred during the Oligocene marine transgression).

Marsupials (p. 70)

Like that of the monotremes, the absence of marsupials from New Zealand has been a long-standing zoological enigma, accentuated by the fact that our nearest neighbour, Australia, has a rich and diverse marsupial fauna. However, unlike that of the monotremes, the marsupial fossil record is somewhat better and has been recently enhanced by discoveries in West Antarctica and North Africa (Woodburne & Zinsmeister 1984; Mahboubi et al. 1983).

Although there is consensus that marsupials originated from ancestral mammalian stocks sometime in the Cretaceous, the probable site of origin is in dispute. Cox (1973) and Tedford (1974), for example, favour South American origins, whereas Kirsch (1984) favours derivation from Australian sources. However, the most commonly adopted scenario is that the marsu-pials originated in middle Cretaceous times, probably in North America, and dispersed from there to South America and in the late Creta-ceous to Australia via Antarctica (Clemens 1979b; Rich 1982; Strahan 1983; Archer 1984b; Woodburne & Zinsmeister 1984). This viewpoint has been adopted in this book (e.g., Fig. 46).

The comparisons of the reproductive systems of monotremes, marsupials, and placentals referred to in the text (p. 70) are for illustrative purposes only, and there is no intention to imply evolu-tionary progression. Modern workers consider both the monotremes and marsupials to be specialised adaptations to their environments and not necessarily "primitive" (e.g., Archer 1984; Augee 1984; Kirsch 1984; Clemens 1979a, b; Lille-graven 1979).

Contribution of biochemical techniques to evolutionary studies

(pp. 58, 76, 90)

Analyses of body proteins and genetic material (DNA) have provided new data on the relation-ships between organisms, as well as the order in which groups appear along specific evolutionary lineages (as all the groups sharing a common lineage share derivatives of certain body proteins and genetic material). The analyses are also capable of mathematical treatment to provide a' timetable for each evolutionary step, providing scientists with a so-called "molecular clock".

The evolutionary relationships between humans and apes and the evolutionary histories of birds and bats are two topics dealt with in this book that have been recently studied by the new bio-chemical techniques. Although such studies have helped clarify relationships within lineages, the dates assigned to the various evolutionary steps have often diverged significantly from previously accepted estimates; a situation that has led to much discussion. Rational discussion has not been helped, however, by the as yet unresolved differences in results from various experimental techniques, and doubts about their mathematical basis (Lewin 1983; Schwartz 1984).

Biochemical studies done so far on hominid evolutionary lineages indicate divergence between the human evolutionary line and that of the African apes occurred either 4–5 million years, or 7–10 million years ago, depending on the biochemical technique used. This contrasts with the commonly accepted date of 15 million years ago (Lewin 1983; Archer & Aplin 1984) (p. 90).

Biochemical analyses of ratites, although support-ing the concept that the ancestors of these flightless birds made their way across Gondwana-land before it broke-up (Fig. 40), also indicate the ancestral stock of the kiwi split off from that of the emu and cassowary at a time (40–45 million years ago) much later than that indicated by geological evidence. The commonly accepted view (e.g., Rich & Balouet 1984), as summarised on p. 58, is that the ancestors of the moa and kiwi walked into New Zealand about 150–140 million years ago, at a time when (judging from all the geological evidence) land connections and climatic conditions in the Southwest Pacific were most favourable for dispersal of land creatures. The biochemical evidence indicates, however, that 40–45 million years ago may be a more likely date for the arrival of ratites in New Zealand. If this date is confirmed by future work the implication is that ancestral ratites reached New Zealand from Australia after the opening of the Tasman Sea, probably via intervening island arcs or emergent land, presumably along the Norfolk and Lord Howe submarine ridges (Diamond 1983; Sibley & Ahlquist 1981, 1983a).

The same biochemical techniques have been used to confirm the antiquity of the New Zealand wren lineage (p. 76). According to Sibley & Ahlquist (1983b) the New Zealand wrens have been isolated from their original ancestral sources for 85–90 million years.

The great Cretaceous die-out

(pp. 72–74)

The hypothesis that the extinctions at the end of the Cretaceous period were caused by the effects of a giant extraterrestrial object striking the earth has been the subject of much debate. Many of the issues involved were highlighted at a major conference on the topic (Silver & Schultz 1982), but wide-ranging discussion has continued in the scientific literature and numerous papers have recently appeared, both in support of a violent cosmic catastrophe (e.g., Bohor et al. 1984; Montanari et al. 1983), and of alternative explanations. The alternative explanations include the effects of a mammoth volcanic explosion, bringing up iridium-laden volcanic debris from the earth's mantle; a similar volcanic eruption on the moon; major climatic changes, related to the break up of Gondwanaland and the movement of continents towards their present positions. There is also debate whether the concentrations of iridium are in fact unique and confined to a single layer at the Cretaceous/Paleocene boundary and also whether the various extinctions of marine and terrestrial organisms were synchronous with each other and with the iridium anomaly (Officer & Drake 1983; Charig & Halstead 1983). Only time and a lot more research will resolve these matters.

Origins and affinities of New Zealand bat fauna (p. 76)

Although the paucity of the fossil record has restricted discussion of the evolutionary patterns of bats, reviews by Ziegler (1982), Hall (1984), and Hand (1984) indicate that early radiation of bats probably occurred in the tropics of Africa, India and Southeast Asia early in the Cenozoic

and that soon after this ancestral bats probably flew to Australia from Southeast Asia. Although there is evidence to indicate that the endemic New Zealand bat, *Mystacina*, may have followed a similar route (notably the presence of a tick, normally parasitic on Australian bats; see Daniel 1979) there is also biochemical evidence indicating links with South America (Pierson et al. 1982; Hand 1984). Both possibilities have therefore been considered (on p. 76).

Date of separation of Antarctica from Australia (pp. 76–78)

The commonly accepted age of 55 million years ago for the start of sea-floor spreading between Antarctica and Australia (Weissel & Hayes 1972) has been challenged by Cande & Mutter (1982), and dated as beginning sometime between 110 and 90 million years. On the other hand, the date of 55 million years has been supported by other recent papers (e.g., Irving 1983; Ferguson et al. 1983). If the dating proposed by Cande & Mutter eventually gains wide acceptance, it should not be assumed that complete separation between Antarctica and Australia necessarily began 110–90 million years ago. This is because Cande & Mutter regard the initial spreading rate between Australia and Antarctica to have been very slow, leaving open the possibility that interchange of land organisms between the two continents may have been possible for some time long after the onset of breakup.

Age of the Pliocene/Pleistocene boundary

In this book (p. 102) the age of the Pliocene/Pleistocene boundary has been "rounded off" to 2 million years because at the time the text was written the boundary was thought to lie somewhere between 1.87 and 1.79 million years (Suggate et al. 1978). However, there are now indications that the age of the boundary may lie closer to 1.63 million years (Beu & Edwards 1984).

Nonetheless, there is no certainty that this date will be universally accepted as marking the start of the Ice Age because the first climatic fluctuations that can be considered to be of glacial/interglacial type began very much earlier, between 3.2 and 2.4 million years ago (e.g., Berggren 1981; Thunell & Williams 1983).

Prehistoric man and faunal extinctions in New Zealand (p. 108)

The record of pre-European extinction of elements within the New Zealand endemic fauna (e.g., the moa, New Zealand swan, New Zealand goose, etc.,) has been reviewed by Cassels (1984). The record indicates that hunting by humans was a prime cause of the catastrophic extermination of fauna that occurred in pre-historic times.

Reconstruction of the New Zealand eagle (*Harpagornis*) (Fig. 65)

The reconstruction of the New Zealand eagle depicted in Fig. 64 is based on a painting by R. J. Jacobs (Canterbury Museum), published in Duff (1949) and Falla (1974). On the basis of new calculations, however, D. H. Brathwaite has proposed that the New Zealand eagle was a short-winged long-tailed forest eagle, with large legs. Although Brathwaite's observations have not as yet been published, a drawing based on his research is included in McCulloch (1982, p. 27).

References

Archer, M. 1984a: Origins and early radiations of mammals. Pp. 477–515 *in* Archer, M.; Clayton, G. *ed.*, Vertebrate zoogeography and evolution in Australasia. Perth, Hesperian Press.
———— 1984b: Origins and early radiations of marsupials. Pp. 631–640 *in* Archer, M.; Clayton, G. *ed.*, Vertebrate zoogeography and evolution in Australasia. Perth, Hesperian Press.

Archer, M.; Aplin, K. 1984: Humans among primates: stark naked in a crowd. Pp. 949–993 in Archer, M.; Clayton, G. ed., Vertebrate zoogeography and evolution in Australasia. Perth, Hesperian Press.

Augee, M. 1984: Quills and bills: The curious problem of Monotreme zoogeography. Pp. 567–570 in Archer, M.; Clayton, G. ed., Vertebrate zoogeography and evolution in Australasia. Perth, Hesperian Press.

Barron, E. J.; Harrison, C. G. A.; Sloan, J. L.; Hay, W. W. 1981: Paleogeography 180 million years ago to the present. Eclogae geologicae Helvetiae 74 : 443–470.

Berggren, W. A. 1981: Role of ocean gateways in climate changes. Stockholm contributions in geology 37 : 9–20.

Beu, A. G.; Edwards, A. R. 1984: New Zealand Pleistocene and late Pliocene glacio-eustatic cycles. Palaeogeography, palaeoclimatology, palaeoecology 46 : 119–142.

Bohor, B. F. et al. 1984: Mineralogic evidence for an impact event at the Cretaceous–Tertiary boundary. Science 224 : 867–869.

Cande, S. C.; Mutter, J. C. 1982: A revised identification of the oldest sea-floor spreading anomalies between Australia and Antarctica. Earth and planetary science letters 58 : 151–160.

Carey, S. W. 1976: The expanding earth. Amsterdam, Elsevier.

Cassels, R. 1974: The role of prehistoric man in the faunal extinctions of New Zealand and other Pacific islands. Pp. 741–767 in Martin, P. S.; Klein, R. G. ed., Quaternary extinctions. A prehistoric revolution. Tuscon, University of Arizona Press.

Charig, A.; Halstead L. B. 1983: Sun sets on the great dinosaur disaster. New Scientist 99 : 606.

Clemens, W. A. 1979a: Notes on the monotremata. Pp. 309–311 in Lillegraven, J. A.; Kielan-Jaworowska, Z.; Clemens, W. A. ed., Mesozoic mammals: the first two-thirds of mammalian history. Berkeley and Los Angeles, University of California Press.

———— 1979b: Marsupialia. Pp. 193–218 in Lillegraven, J. A.; Kielan-Jaworowska, Z.; Clemens, W. A. ed., Mesozoic mammals: the first two-thirds of mammalian history. Berkeley and Los Angeles, University of California Press.

Cohee, G. V.; Glaessner, M. F.; Hedberg, H. D. 1978: Contributions to the geologic time scale. American Association of Petroleum Geologists, studies in geology 6.

Cox, C. B. 1973: Systematics and plate tectonics in the spread of marsupials. Pp. 113–119 in Hughes, N. F. ed., Organisms and continents through time. Palaeontological Association, special papers in palaeontology 12.

Daniel, M. J. 1979: The New Zealand short-tailed bat, Mystacina tuberculata; a review of present knowledge. New Zealand journal of zoology 6 : 357–370.

Diamond, J. M. 1983: Taxonomy by nucleotides. Nature 305 : 17–18.

Duff, R. 1949: Pyramid valley: the story of New Zealand's greatest moa swamp. Christchurch, Canterbury Museum.

Falla, R. A. 1974: The moa. New Zealand's nature heritage 1(3) : 69–74.

Ferguson, K. U.; Kelly, P. R.; Gleadow, A. J. W.; Lovering, J. F. 1983: Australia–Antarctic break-up: apatite fission track evidence from South and Western Australia and East Antarctica. P. 582 in Oliver, R. L.; James, P. R.; Jago, J. B. ed., Antarctic earth science. Canberra, Australian Academy of Science.

Firstbrook, P. L.; Funnell, B. M.; Hurley, A. M.; Smith, A. G. 1979: Paleoceanic reconstructions 160–0 Ma. Washington, National Science Foundation.

Griffiths, M. 1983: Lactation in monotremata and speculations concering the nature of lactation in Cretaceous multituberculata. Acta palaeontologica Polonica 28 : 93–102.

Grindley, G. W.; Davey, F. J. 1982: The reconstruction of New Zealand, Australia and Antarctia (review). Pp. 15–29 in Craddock, C. ed., Antarctic geoscience. Madison, Wisconsin, University of Wisconsin Press.

Grindley, G. W.; Oliver, P. J. 1983: Paleomagnetism of Cretaceous volcanic rocks from Marie Byrd Land, Antarctica. Pp. 573–578 in Oliver, R. L.; James, P. R.; Jago, J. B. ed., Antarctic earth science. Canberra, Australian Academy of Science.

Hall, L. 1984: And then there were bats. Pp. 837–852 in Archer, M.; Clayton, G. ed., Vertebrate zoogeography and evolution in Australasia. Perth, Hesperian Press.

Hand, S. 1984: Bat beginnings and biogeography: a southern perspective. Pp. 853–904 in Archer, M.; Clayton, G. ed., Vertebrate zoogeography and evolution in Australasia. Perth, Hesperian Press.

Harland, W. B.; Cox, A. V.; Llewellyn, P. G.; Pickton, C. A. G.; Smith, A. G.; Walters, R. 1982: A geologic time scale. Cambridge, Cambridge University Press.

Howell, D. G. 1980: Mesozoic accretion of exotic terranes along the New Zealand segment of Gondwanaland. Geology 8 : 487–491.

Howell, D. G.; Jones, D. L.; Cox, A.; Nur, A. ed. 1984: Proceedings of the circum-Pacific terrane conference. Stanford University publications in the geological sciences 18.

Irving, E. 1983: Fragmentation and assembly of the continents, mid Carboniferous to present. Geophysical surveys 5 : 299–333.

Kamp, P. J. J. 1980: Pacifica and New Zealand: proposed eastern elements in Gondwanaland's history. Nature 288 : 659–664.

———— 1984: Neogene and Quaternary extent and geometry of the subducted Pacific Plate beneath North Island, New Zealand: implications for Kaikoura tectonics. Tectonophysics 108 : 241–266.

Kemp, T. S. 1982: Mammal-like reptiles and the origin of mammals. London, Academic Press.

———— 1983: The relationships of mammals. Zoological journal of the Linnean Society of London 77 : 353–384.

King, L. C. 1983: Wandering continents and spreading sea floors on the expanding earth. New York, John Wiley.

Kirsch, J. 1984: Marsupial origins: taxonomic and biological considerations. Pp. 627–631 in Archer, M.; Clayton, G. ed., Vertebrate zoogeography and evolution in Australasia. Perth, Hesperian Press.

Lewin, R. 1983: Is the Orangutan a living fossil? Science 222 : 1222–1223.

Lillegraven, J. A. 1979: Reproduction in Mesozoic mammals. Pp. 259–276 in Lillegraven, J. A.; Kielan-Jaworowska, Z.; Clemens, W. A. ed., Mesozoic mammals: the first two-thirds of mammalian history. Berkeley and Los Angeles, University of California Press.

Lock, R. G. 1983: Continental margin petroleum potential in the Ross Sea region. New Zealand Antarctic record 5(1) : 6–15.

Mahboubi, M.; Ameur, R.; Crochet, J.-Y.; Jaeger, J.-J. 1983: Comptes rendus de l'Academie des Sciences Paris 297 Ser. C, serie II: 691–694.

McCulloch, B. 1982: No moa: some thoughts on the life and death of New Zealand's most spectacular bird. Christchurch, Canterbury Museum.

McGowan, C. 1982: The wing musculature of the Brown Kiwi Apteryx australis mantelli and its bearing on ratite affinities. Journal of zoology 197 : 173–219.

———— 1984: Evolutionary relationships of ratites and carinates: evidence from ontogeny of the tarsus. *Nature 307*: 733–735.

Montanari, A. et al. 1983: Spheroids at the Cretaceous-Tertiary Boundary are altered impact droplets of basaltic composition. *Geology 11*: 668–671.

Murray, P. 1984: Furry egg-layers: the monotreme radiation. Pp. 571–626 in Archer, M.; Clayton, G. *ed.*, Vertebrate zoogeography and evolution in Australasia. Perth, Hesperian Press.

Nur, A.; Ben-Avraham, Z. 1977: Lost Pacifica continent. *Nature 270*: 41–43.

Odin, G. S. *ed.* 1982a: Numerical dating in stratigraphy. New York, John Wiley.

———— 1982b: The Phanerozoic time scale revisited. *Episodes 1982 (3)*: 3–9.

———— et al. 1983: Numerical dating of Precambrian–Cambrian boundary. *Nature 301*: 21–23.

Officer, C. B.; Drake, C. L. 1983: The Cretaceous-Tertiary transition. *Science 219*: 1383–1390.

Owen, H. G. 1976: Continental displacement and expansion of the Earth during the Mesozoic and Cenozoic. *Philosophical transactions Royal Society London A281*: 229–291.

———— 1981: Constant dimensions or an expanding earth? Pp. 179–192 in Cocks, L. R. M. *ed.*, The evolving earth. London, British Museum (Natural History) and Cambridge University Press.

———— 1983a: Ocean-floor spreading evidence of global expansion. Pp. 31–58 in Carey, S. W. *ed.*, The expanding earth: a symposium. Hobart, University of Tasmania.

———— 1983b: Some principles of physical palaeogeography. Pp. 85–114 in Sims, R. W.; Price, J. H.; Whalley, P. E. S. *ed.*, Evolution, time and space: the emergence of the biosphere. London, Academic Press.

———— 1983c: Atlas of continental displacement 200 Ma to present. Cambridge, Cambridge University Press.

Palmer, A. R. *comp.* 1983: The decade of North American geology 1983 geologic time scale. *Geology 11*: 503–504.

Parrish, J. T.; Ziegler, A. M.; Scotese, C. R. 1982: Rainfall patterns and the distribution of coals and evaporites in the Mesozoic and Cenozoic. *Palaeogeography, palaeoclimatology, Palaeoecology 40*: 67–101.

Pierson, E. D.; Sarich, V. M.; Lowenstein, J. M.; Daniel, M. J. 1982: Mystacina is a Phyllostomatoid Bat. *Bat research news 23(4): 78*.

Retallack, G. J. 1984: Origin of the Torlesse terrane and coeval rocks, South Island, New Zealand: discussion and reply. *Bulletin Geological Society of America 95*: 980–982.

Rich, P. V.; Balouet, J. 1984: The waifs and strays of the bird world or the ratite problem revisited, one more time. Pp. 447–455 in Archer, M.; Clayton, G. *ed.*, Vertebrate zoogeography and evolution in Australasia. Perth, Hesperian Press.

Rich, T. H. 1982: Monotremes, placentals and marsupials: their record in Australia and its biases. Pp. 385–478 in Rich, P. V.; Thompson, E. M. *ed.*, The fossil vertebrate record of Australasia. Melbourne, Monash University Press.

Schwartz, J. H. 1984: The evolutionary relationships of man and orang-utans. *Nature 308*: 501–505.

Scotese, C. R.; Bambach, R. K.; Barton, C.; van der Voo, R.; Ziegler, A. M. 1979: Paleozoic base maps. *Journal of geology 87*: 217–277.

Shields, O. 1983: Trans-Pacific biotic links that suggest earth expansion. Pp. 199–205 in Carey, S. W. *ed.*, Expanding Earth Symposium, 1981. Hobart, University of Tasmania Press.

Sibley, C. G.; Ahlquist, J. E. 1981: The phylogeny and relationships of the ratite birds as indicated by DNA–DNA hybridization. Pp. 301–335 in Scudder, G. G. E.; Revea, J. L. *ed.*, Evolution today. *Proceedings of the 2nd International Congress on Systematics and Evolutionary Biology*.

———— 1983a: The phylogeny and relationships of the ratite birds as indicated by DNA–DNA hybridization. *Abstracts, 15th Congress, Pacific Science Association. p. 213*.

———— 1983b: The phylogeny of the passerine birds of Australia and New Zealand, based on DNA–DNA hybridization data. *Abstracts, 15th Congress, Pacific Science Association. p. 213*.

Silver, L. T.; Schultz, P. H. *ed.* 1982: Geological implications of impacts of large asteroids and comets on the earth. *Geological Society of America special paper 190*.

Smith, A. G. 1981: Phanerozoic equal-area maps. *Geologische rundschau 70*: 91–127.

Smith, A. G.; Briden, J. C. 1977: Mesozoic and Cenozoic paleocontinental maps. Cambridge, Cambridge University Press.

Smith, A. G.; Hurley, A. M.; Briden, J. C. 1981: Phanerozoic palaeocontinental world maps. Cambridge, Cambridge University Press.

Smith, D. G. 1982: Historical geology: layers of earth history. Pp. 386–409 in Smith, D. G. *ed.*, The Cambridge encyclopedia of earth sciences. Cambridge, Cambridge University Press.

Strahan, R. *ed.* 1983: The Australian Museum complete book of Australian mammals. Melbourne, Angus & Robertson.

Stevens, G. R. 1980: Geological time scale. Wellington, Geological Society of New Zealand.

Suggate, R. P.; Stevens, G. R.; Te Punga, M. T. *ed.* 1978: The geology of New Zealand. Wellington, Government Printer.

Tedford, R. H. 1974: Marsupials and the new paleogeography. Pp. 109–126 in Ross, C. A. *ed.*, Paleogeographic provinces and provinciality. *Society of Economic Paleontologists and Mineralogists special publication 21*.

Thunell, R. C.; Williams, D. F. 1983: The stepwise development of Pliocene–Pleistocene paleoclimatic and paleoceanographic conditions in the Mediterranean: oxygen isotopic studies of DSDP sites 125 and 132. *Utrecht micropaleontological bulletins 30*: 111–127.

Tozer, E. T. 1982: Marine Triassic faunas of North America: their significance for assessing plate and terrane movements. *Geologische rundschau 71*: 1077–1104.

———— 1984: The Trias and its ammonoids: the evolution of a time scale. *Geological Survey of Canada miscellaneous report 35*.

Walcott, R. I. 1984: Reconstructions of the New Zealand region for the Neogene. *Palaeogeography, palaeoclimatology, palaeoecology 46*: 217–231.

Weissel, J. K.; Hayes, D. E. 1972: Magnetic anomalies in the Southeast Indian Ocean. Pp. 165–196 in Hayes, D. E. *ed.*, Antarctic oceanology II. The Australian–New Zealand sector. *American Geophysical Union Antarctic research series 19*.

Woodburne, M. O.; Zinsmeister, W. J. 1984: The first land mammal from Antarctica and its biogeographic implications. *Journal of paleontology 58*: 913–948.

Ziegler, A. C. 1982: An ecological check-list of New Guinea recent mammals. Pp. 863–894 in Gressitt, J. L. *ed.*, Biogeography and ecology of New Guinea. volume 2. *Monographiae biologicae 42*. The Hague, Dr W. Junk.

Index

BY AUTHORITY: V. R. WARD, GOVERNMENT PRINTER, WELLINGTON, NEW ZEALAND—1985

20525B—85PTK

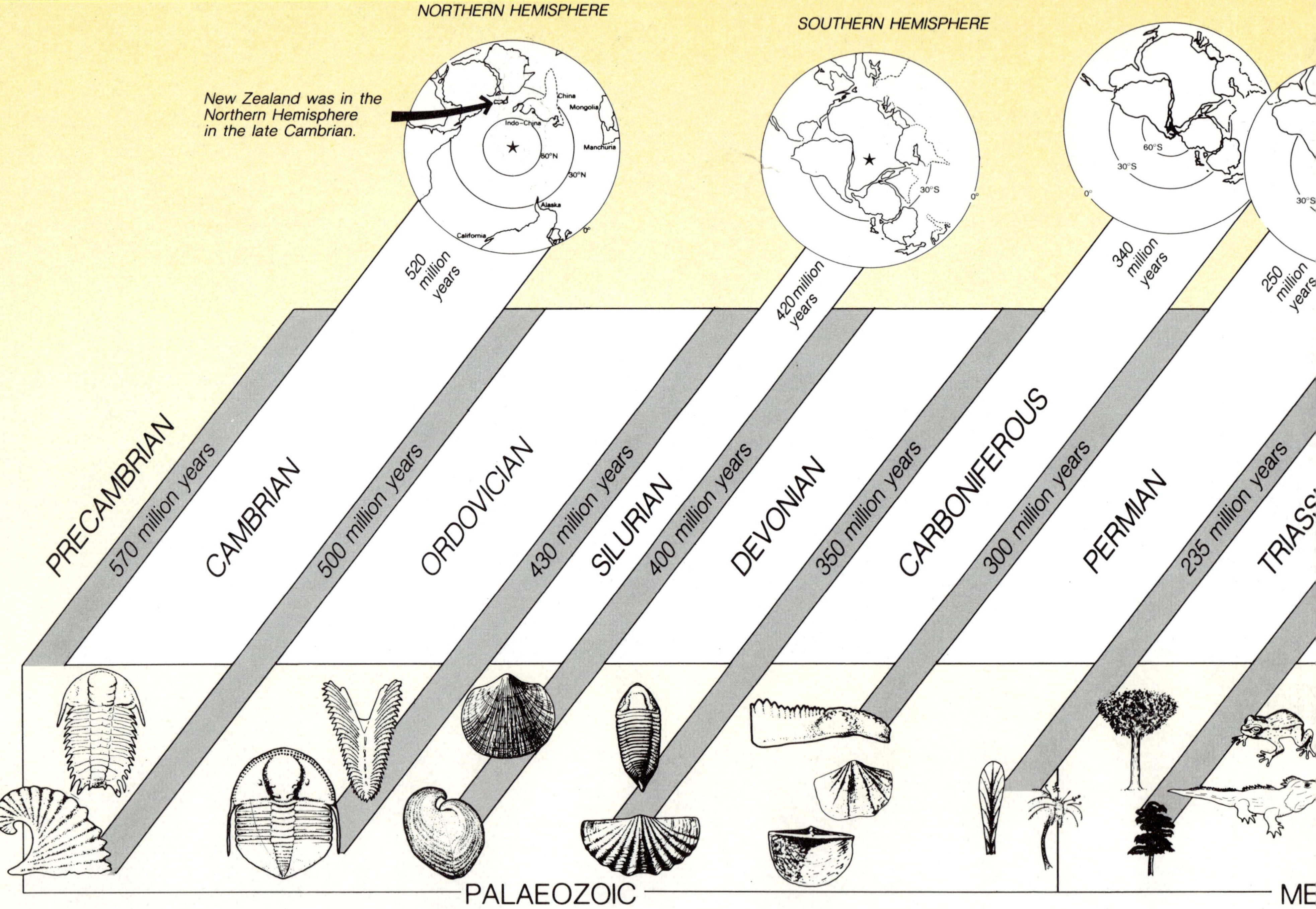

NORTHERN HEMISPHERE
SOUTHERN HEMISPHERE
New Zealand was in the Northern Hemisphere in the late Cambrian.
China
Mongolia
Indo-China
Manchuria
60°N
30°N
Alaska
California
0°
520 million years
420 million years
340 million years
250 million years
30°S
0°
60°S
30°S
0°
30°S
PRECAMBRIAN
570 million years
CAMBRIAN
500 million years
ORDOVICIAN
430 million years
SILURIAN
400 million years
DEVONIAN
350 million years
CARBONIFEROUS
300 million years
PERMIAN
235 million years
TRIASS
PALAEOZOIC
ME